# Rock of Ages

A Cosmic Love Story

Dylan;
For the past, presant
and future.
Sept. 9-2015
Mark Brown

Library and Archives Canada Cataloguing in Publication
Pogue, Carolyn 1948 -
Rock of Ages: A Cosmic Love Story

Includes bibliographical references and index.

QE475.G55P64 2013 552'.4 C2013-903688-1

1. ecology 2. geology 3. gneiss —Northwest Territories — Acasta River Region 4.Title
5. mythology 6. spirituality 8. narrative nonfiction 8. Title 9. Carolyn Pogue

Cover photo of the Acasta River provided by Dr. Wouter Bleeker, Research Scientist at the Geological Survey of Canada, Ottawa.

Back cover photo by Jeremy Emerson Author photo by Bill Phipps
Inside photos by Creative Commons; photographers listed on page 105.
Map by Brock University.

First printing August, 2013 under the title Rock of Ages: The oldest known rock on Earth, and then some...
Second printing Rock of Ages: A Cosmic Love Story March 2015



Please order paper books through your favorite book store or send order to playingforlife@shaw.ca Please order ebooks through www.rockofagesnwt.com

Playing for Life Publishing
918 Seventeen Street NW Calgary,
Alberta T2N 2E4

Printed by Blitz Print Inc., Calgary, Canada using recycled, nonchlorinated paper

ISBN 978-0-9733284-2-4

# Rock of Ages

A Cosmic Love Story

Carolyn Pogue

Playing for Life

Raven

## Other books by Carolyn Pogue

### For children

After the Beginning
A Creation Story
Counting Peace
The Colour of Peace

### For young adults

West Wind Calling
Gwen
World of Faith: Introducing Spiritual Traditions to Teens

### For adults

Language of the Heart

### For teachers and youth leaders

A New Day: Peacemaking Stories and Activities
Remember Peace
Seasons of Peace

# Dedication

**For Children**

# Contents

When I first arrived in Fort Chipewyan [northern Alberta], the locals pressed me: Why did I want to see the north shore of Lake Athabasca? Why the north shore with its sheer cliffs, frequent storms, difficult, even dangerous waters? Why not the south shore, all sand and beach? They knew cabins, people all along, not far from town, easy access by boat. "I want to see the rock," I said simply. The Canadian Shield is some of the oldest rock in the world.

When they still looked puzzled, wanting to find the common language between us, I managed finally, "It's the closest thing to the beginning of things. It's the closest thing to the Creator." Then they nodded in recognition.

Audrey J. Whitson in *Teaching Places*

# Acknowledgements

**D**eep thanks to Bill Phipps who supports every project I dream up with grace, generosity and love. Thanks to my nephew Mark Brown who first gave me a piece of the Acasta gneiss and who has generously provided stories, information and inspiration for this book.

Thank you to each patient person who agreed to an interview: Dr. Janet King, Walter Humphries, Jack Walker, Jeremy Emerson, Healer Mary Botel, Elder Doreen Spence, Dr. Wouter Bleeker, Cultural Teacher Fred Sangris, Dr. Gina Marie Ceylan and Dr. Tom Feuchtwanger. Thank you also to Aggie Brockman, Andrea Czarnecki, Marie Saretsky, Jennifer Day and Gail Sidonie Sobat. For friends along the way who asked questions or offered insights about their personal connections to rocks, thank you.

Thank you to each boy and girl, woman and man who works hard to protect and replenish our fragile planet.

Mountain Avens

# Introduction

**The one who sat on the ground in the tipi meditating on life and its meaning, accepting the kinship of all creatures and acknowledging unity with the universe, was infusing into his being the true essence of civilization.**

Chief Luther Standing Bear, Oglala Sioux

I have always felt connected to Earth, but the story of one particular rock has connected me to her in a new way. My preparation to write about the oldest known rock on Earth likely began forty years ago when I lived in Yellowknife, Northwest Territories, and specifically when I was invited in to the Circle. That first time I sat in the sacred Circle was profound.

I had driven the dusty 100 km from Yellowknife to Behchoko, known in those days as Fort Rae, with René Fumoleau, a photographer-priest-historian. The autumn air was crisp, above us the Northern Lights danced. Change was in the air.

We sat with people I'd never met, mostly Tlicho First Nation, then referred to as Dogrib Indians. It was an early experience of being a minority race for me; I am Caucasian. The invited Elder was Eddie Bellerose, a Cree from Alberta. Using the Medicine Wheel, Eddie gave a teaching about the four directions. The wheel was large, and he created it on the floor to demonstrate as he spoke. He used other tools to explain his worldview, too, stories and humour. I was a relative newcomer to the North. I had not yet learned much about the Dene, Treaty 11 or the Precambrian land I'd come to. And yet, sitting in that Circle, I felt deeply as if I'd arrived home.

Over the years since that day, I have had the privilege of sitting in other sacred Circles in Alberta, Manitoba, Ontario and in India and Zimbabwe. I have always had the same feeling of being *home.* In women's sacred Circles led jointly by Asiniwaciskwiw/Lorraine Sinclair, a Cree cultural teacher, and her friend Celtic storyteller Andrine Morse, I came to believe that it was the Circle that drew me in, and resounded in my cells and bones.

The Circle anchors me in my spirituality and in my worldview. At one time, of course, all of our ancestors sat in the sacred Circle for storytelling, teaching, celebrations and ceremonies.

That first Circle, far from the part of Canada where I'd grown up, opened a door for me to understand myself and regain pieces of my spirit that had been confused by straight lines and patriarchy.

I lived in Yellowknife, Northwest Territories, or *Sombe K'e, Denendeh* in the language of the people, from 1970 until 1986. This was in the olden days, before Ice Road Truckers, Arctic Air or North of Sixty were aired on CBC television. It was before children in Canadian schools learned the real history of First Nations.

To many, the North then was about white explorers arriving from far away and "discovering" things; a mysterious place, often portrayed as dangerous, exotic, wild. It was so mysterious that educators did not imagine that residential schools were harmful or that culturally appropriate books might help kids fare better in the school system. In those days little kids who skinned rabbits, set snares and fish nets were learning to read English from *Dick and Jane* books. Father wore a fedora and mother, in high heels, stood by the white picket fence as the children, Puff the cat and Spot, the useless dog, watched.

I loved Yellowknife. I had thought it would be fun to live there for a year or so, but like so many others, that year stretched on and on. The times were exciting. These were the years of new oil discoveries and of the Berger Inquiry into how a pipeline down the MacKenzie Valley would affect communities and the land. The Dene Nation was born; the Tree of Peace Friendship Centre first opened. CBC began broadcasting in Dene languages. The Territorial government changed from being run by white males appointed by Ottawa to a government of the people. Leaders like Cindy Kenney Gilday, Francois Paulette and Georges Erasmus began the long process of educating southern Canadians and indeed the world about their land and culture. Rapid change was the order of the day for Canada, the territory and me.

Our family home was on Latham Island, in Old Town, at the north end of Yellowknife. Our living room window looked out on Back Bay where in summer we watched planes, canoes and cabin cruisers, and in winter, we enjoyed seeing dog teams, skidoos, skied planes and cross country skiers on the ice. Although life on the bay is much the same today (with the notable exception of colourful houseboats), the Old Town has become somewhat gentrified.

I knew that I lived — literally — over two gold mines. Once in a while we felt the earth beneath our home tremble when Giant Mine blasted deep underground. But I was ignorant about the land I lived on and didn't even know the age of the rocks I looked at every day.

Although I've always picked up interesting rocks, admired them as works of art, I didn't really know what I was looking at. I was acquainted with some geologists, but in those years I was preoccupied with my young family, community, church, theatre and peace work. My first book, *Yellowknife,* was published there and I began writing in earnest. I worked at Mildred Hall Elementary School. Needless to say, I was busy.

My understanding of geology was superficial and minimal. The one time I travelled to a tundra mining camp I was more interested in looking for birds' nests than in learning about minerals beneath my feet. Around the time I left the North, the Acasta gneiss was discovered by Dr. Janet King and a team from the Geological Survey of Canada. If I heard of this, it did not register.

I learned of the Acasta gneiss (pronounce "nice") in 2010 when my nephew gave me a piece. My husband and I have hauled stones home from various oceans, rivers, deserts and lake shores around the world, so I was happy to receive it. It's a pretty rock with lines, squiggles and sparkles in it. Instinctively I kept it apart from the others; this one was different.

From time to time I'd look at it, ask friends what they thought about a rock more than 4 billion years old. I'd never thought much about the age of other rocks that live in our home and garden. This one made me realize how little I knew about the geologic history of my planet home.

*Rock of Ages* is different from any book I have written or edited. Its research has taken me down wild new trails of geology, mythology, biology, theology and ecology. It has led me to interview Elders, claim holders, geologists, Reiki masters and prospectors. What a privilege.

For me, this rock has underlined the truth of Chief Seattle's teaching that we are but one strand in the web of life. It has underlined the truth of ancient stories handed down over generations. I've learned some scientific theories and discoveries, and feel connected anew to our planet home. It has helped me know why I felt so at home, sitting in that first Circle so long ago.

The original name of the river in Tlicho is Denàdzìideè. According to Georgina Chocolate of the Tlicho government this translates roughly as "river-chased-river."

In the language of the people, *Mussi Cho* for joining me on the pages of this book.

Carolyn Pogue
Calgary, Alberta

Chapter 1

# Once Upon a Time

**God made time.... and plenty of it.**

Mrs. O'Malley, Ireland

The Acasta gneiss is more than four billion years old. Its story is older than any other on the planet. I've always thought of Earth as our Mother and have marvelled at the wonder of evolution, but after pondering this gneiss I thought, "Forget about apes! Forget even about wiggly things emerging from the water. This *rock* is our original ancestor." And that was before I learned the Lakota Sioux Creation story.

The Lakota Sioux say that in the beginning, everything was in the mind of the Great Spirit, *Wakan-Tanka*. And everything that was to be, already existed in spirit. These spirits wanted to *become.* They travelled the whole universe, looking for a place to manifest themselves. Finally they arrived on Earth. But the spirits needed land, and the planet was only water at that time. And so they caused a great rock to rise up from the water. This was Grandfather Rock, the first one. After some time, amidst steamy clouds, more land appeared.

The story teaches that rocks must be respected because they are the beginning of all Creation. People are helped to remember this when they enter the sweat lodge and see water vapour rising from hot stones.

The world is filled with ancient stories passed through the generations to us. From the time humans could tell stories, it seems, we've been trying to understand where we came from, how it all began.

Perhaps you grew up with the story of a Spirit God who created the world in six days, separating light from dark, land from water. Or maybe you were fed, along with your milk, the story of Lord Vishnu awakening at the dawn of time, floating on an endless sea. Or maybe you learned about P'an Ku, hatched from a cosmic egg. Possibly your parents told you about Chaos, the gaping emptiness, from which emerged Gaia. Creation stories are filled with wonder and reverence and contain lessons about respect.

Stories shape us. Whether they are scientific or mythological, they help to mold our relationship with Nature. Generally, it seems, western religious myths set us apart from and above Nature, while Indigenous and eastern traditions see humankind as an integral part of Nature. But before formal religions developed, our ancestors told similar stories around the world.

No matter where they lived, the storytellers gathered listeners around a fire and began. On mountains, under trees, in caves and tents and huts they began their stories roughly the same way.

"In the beginning," they said, each in their own language, "this is how the world was born." And they described the world coming to birth long before humankind. I imagine the listeners looking up to the wide night sky, seeing the Northern Lights, twinkling stars or the silver moon. Wonder and awe fill them. When they felt the good Earth beneath them, they knew they were connected to her.

In those earliest times, we humans knew intimately that we could not survive without Earth's generosity. We wanted to please her. Some carved small stone goddesses to show reverence. We worshipped by gigantic standing stones under the moon. We created circular worship places of stone on or under the ground. We didn't understand all the science of our planet, but simply knew that it was awe-some.

Our earliest Creation stories were about fire and wild wind, floods, volcanoes, clamshells and spiders. They were of star dust, gods and goddesses who brought this world into being for love. The stories were born in the richness of the human heart, in reverence and wonder.

And then in the 1600s a new tribe of storytellers added their tales of wonder. Scientists. They began categorizing bits and pieces of the planet, naming, dissecting, stuffing and preserving in boxes, under glass, in museums. By the Victorian era, we knew we were quite clever about manipulating, medicating, managing and mapping our world. *Taming* it, we thought. The world would serve us. Our appetite for control was, and is, insatiable. The scientific discoveries came fast and furious. They still do. Artificial insemination, petrie dishes, cloning, the genetic code, continental drift, evolution, dinosaurs and DNA have all been discovered, in geological time, in the blink of an eye. These stories, too, filled listeners with awe. But for many, the awe shifted from the wonder of Creation to the wonder of how well we could exploit and manipulate Creation. Myth makers began to be seen as naive.

Interestingly, the scientific storytellers used much the same language to describe the beginning of the universe. They described chaos, darkness, flaming light and wild energy. They described a world coming to birth long before humankind.

Sitting in laboratories or learning through modern methods often removed us from the feel of the good green Earth. Land was manicured, controlled or buried under pavement. Even seasonal food became a thing of the past for some people, and this further disconnected us from Earth. The night sky, in many parts of the world, became obscured by artificial light. Modern people began to feel disconnected, but we didn't realize that at first. There was a shift from seeing with our hearts, one could say. Even little children became disconnected from the natural world. We were losing touch.

Gordon Hampton is an acoustic ecologist, which means he studies the sounds of the natural world. He says, "More than ever before, we need to fall back in love with the land." It's hard to do that if we are disconnected from her. It's difficult to do that if we think we must choose between a sacred or mythological understanding of Earth and a modern, scientific understanding. Or, if we think we must choose between Earth care and a job.

But the science now reveals that when the ancient ones said that life had come from the skies, they were right. Science and myth share the idea too, that humans are indeed created from soil.

Ronald J. Glasser, MD wrote of this in his book *The Body is the Hero.* "The fluids in our bodies mimic the primeval seas in which we began. The concentrations of salts, of sodium, potassium and chloride in our bloodstream, the cobalt, magnesium and zinc in our tissues, are the same as those that existed in the earliest seas.

"Not only does our blood go back to those ancient seas; we are also, literally, children of the earth. The carbon in our bones is the same carbon that forms the rocks of the oldest mountains. The molecules of sugar that flow through our bloodstream once flowed in the sap of now fossilized trees, while the nitrogen that binds together our bones is the same nitrogen that binds the nitrates to the soil. Life has endured as long as it has because it is formed from substances as basic as the earth itself."

Imagine. Your tears could have once been slurped into a dinosaur's mouth; your saliva could have once flowed down the Ganges River or been in an iceberg. How did the ancient storytellers know we were created from the very Earth? How did the biblical writers, the Hindu myth makers the Iroquois holy people imagine our start?

The scientific discoveries and deductions about how the world was created are so recent that they are in living memory. My grandparents, for example, were born around the time that Darwin first published his book, *The Origin of the Species.* When naturalist David Attenborough asked about continental drift in 1947, his professor replied that the idea of a continent moving was "sheer moonshine."

In the latter part of the last century, just when we thought we had it all figured out (why, we had even photographed ourselves from outer space!) we began to understand how little we know. We found new species in the rainforest and in the deep oceans. Our technology allowed new and amazing insights. Literally. In 1984, Canadian scientists found a way to map the inside of our planet in order to see down into the core of its being. They launched the *Lithoprobe* project. This allows scientists to take internal pictures and soundings, rather like the ultrasounds pregnant women have to picture their babies before birth. ("Only," a man told me seriously, "much more complicated!" Of course!)

We had other insights, too. We found out that there is a limit to natural resources. We learned that our over-consumption and disconnection from Nature is stressing us, the oceans and lakes, the air, the soil and other creatures — to death. Some believe we are headed toward disaster. Unless we change. We need to re-view how we live. We need a new story. And we've got one.

The writings of Thomas Berry have inspired a generation. Berry was an American Catholic priest, cultural historian and eco-theologian who helped combine the scientific and the sacred views. Brian Swimme, director of The Center for The Story of the Universe in California carries on Berry's work.

Swimme received his Ph.D from the University of Oregon for work in gravitational dynamics. He applies the context of *story* to help us understand the 13.7 billion year trajectory of the universe. Such a story, he feels, will assist in the emergence of a flourishing Earth Community. "Scientists have discovered what Native peoples have always told us: We are all one," he says.

These days, Swimme travels the world speaking about "the new cosmology," an understanding of the world in the context of the story of the universe. Like his mentor, Thomas Berry, he is an author. *Journey of the Universe,* written with Yale researcher Mary Evelyn Tucker, confirms the notion that somehow, humans have always felt a connection to stars.

They write, "In many cultures throughout history humans intuited that they descended from the stars, even before they had the empirical evidence from science that our bodies were formed by the elements forged by the stars. Humans felt something in the depths of the night as they contemplated the presence of the stars. They began to suspect that the meaning of their lives went far beyond what preoccupied them during the urgencies of the daytime world. They knew in their hearts that their journey and the radiance of the stars were interwoven."

Swimme's enthusiasm for this time of awakening is evident. "We are Earth become conscious!" he says joyfully. "We had become so focused on humans that our relationship with the rest of Nature atrophied. The new cosmology allows us to understand the continuity and to reconnect; to understand that the Earth gave birth to us. We are the universe in the form of a human, related in terms of energy and of matter. We are part of an amazing story!"

The amazing story is part of geologist Gina Marie Ceylan's enthusiasm, too. I had mailed a piece of the Acasta gneiss to her at the University of Missouri, and she held it as we spoke by phone. "Generally as I've become familiar with rocks and minerals, I've got to the point where I can pick up just about any rock and almost immediately know what it is. With this rock, I feel the weight of it relative to its size, its texture, density, the way it's broken and cut, what the surfaces feel like. Sometimes when I'm teaching I get students to describe to me what they see visually and then to describe other things. Initially they have a very hard time; but it tends to be a combination of those different observations, using different senses, that helps them to identify their rock sample. Knowledge informs our perception, and perception feeds back new knowledge.

"When you think that this rock is 4.03 billion years old, then all human civilization, everything we've thought and done, it's just nothing. It's overwhelming to think that, in the grand scheme of things, we're insignificant. At the same time, it brings out an ephemeral kind of experience. Life is valuable because it's so brief.

"Holding this rock reminds me of when I was little, seeing the sunlight sparkling off the ocean. I thought in my little girl brain, 'That's magic sparkling out there.' Knowing what's happening [scientifically] to make water sparkle doesn't take that away. It enhances it, makes it even more beautiful.

"I'm a scientist. I don't say that in any kind of religious or spiritual way — it's *real* magic. That's the first thing that hit me. This rock is magic, too. You can't compare this rock to anything because most of what we know is insignificant in comparison. And that's mind-wrenchingly beautiful.

"And I don't want to call it power, but it's a weight of significance. It tells us a lot of stories especially as we've dated it and learned details of its geochemistry... to learn about the earliest crust [of our planet]. It's important to me that it's a metamorphic rock, a rock that has changed from one thing to another.

"Heat, pressure and fluids have moved through this rock and changed it, but it's still essentially the same rock. It's been through so much; it's analogous to human experience. In life we go through all this *stuff,* good and bad, and it changes us. Especially the more challenging things. We might experience minor stuff that is challenging but often times if there is a huge event or challenge it overprints everything else. After that point, nothing else can touch us. I feel this about this rock. It gives you a different perspective."

Gina Marie knows a lot about perspective. She is a runner, a climber, swimmer, wife and PhD student. And, due to a degenerative condition, she is blind.

This obviously doesn't prevent her seeing what so many cannot see, even with physical sight. Her thesis is in Science Education, her passion is geoscience education and making it more accessible and inclusive. "Classical geology is focussed on what we can see, but it's often misleading and doesn't stand on it's own. You really need a diversity of perceptions to understand as much as possible," she explains.

Gina Marie also takes the long view of life on Earth. "I'm concerned that most humans ... have certain tendencies that are fatal to our race and harmful to our planet. We're over-consuming. We're not sustainable. The planet will recover, but life is fragile and the balance is delicate."

Another American geologist expressed a similar concern in her book, *Reading the Rocks: The Autobiography of the Earth.* Marcia Bjornerud wrote, "What can humans do to leave a more distinguished entry in the geologic record?" She then answered her own question:

"For a start, we can spend more time conversing with existing rocks, whose gravitas may help us stop thinking in throwaway sound bites and anesthetizing euphemisms. Rocks will tell us that we can trust the Earth, that it is immensely old, durable and immensely wiser and more patient than humans.

"Rocks will remind us that arms races never have victors. Rocks may cause us to realize that mountainous garbage dumps, unopposed predation, unchecked consumption, and the flux of commodities from the poor to the rich are violations of ancient laws mandating recycling and redistribution."

She writes, "Rocks may even cause us to rediscover thoughtful discourse about complex environmental issues and to instill in children an appetite for understanding deep origins and histories. Maybe.

"When I despair at the state of things, rocks always offer some comfort. I see gneisses and limestones and granites, greenstones and blueschists and red beds, and I think to myself, what a wonderful world."

I couldn't agree more. To me it is a wonder-filled world, and taking a walk with a three year old is a good reminder of this.

But whether we were shaped by a scientific, mythological or cosmological story, now we know that the Acasta gneiss, at 4.03 billion years of age, is one more story we can tell.

**Questions for Reflection**

2. The author writes that, "Stories shape us." What is your foundational story that connects you to the natural world?
3. When, or if, you despair at the state of the world, what offers you comfort?

Arctic Hare

**Family Album**

4.6 billion years ago, Earth was born
4.3 billion years ago, the Acasta rock emerged
3.4 billion years ago, stromatolites exhaled
500 million years ago, fish swam
400 million years ago, insects and seeds appeared
150 million years ago, birds took to the skies
130 million years ago, flowers blossomed
200 thousand years ago, we showed up.....

**Carbon dating for dummies**

Geologist Tom Feuchtwanger patiently explained a dumbed down version of carbon dating for me. I am grateful.

*Imagine taking an ice cube from the freezer. You note its dimensions and set it on a plate in the kitchen. You note the conditions of the room, such as temperature. You know the rate at which ice melts under various conditions. You leave it on the plate, and return sometime later to measure it, again noting conditions in the room. You would be able to calculate the time elapsed from the time you removed it from the freezer to the present moment by the difference in those measurements.*

*We know the rate at which chemical elements deteriorate and so, working backwards, we can figure out how long something has been on the planet.*

Chapter 2

# In the Beginning. Really.

**It makes sense that Canada would cultivate world-class geologists. Within our vast boundaries, you'll find an example of every type of geology on the planet. From the oldest rocks in the world ... to the youthful Rocky Mountains, and the fossil-bearing badlands, it's a geological candy store.**

Bob MacDonald's foreword to
*Ghost Mountains and Vanished Oceans*

Trying to understand our origins and Earth Science seems to be part of our Canadian DNA. My geologist friend from South Africa argues that "everyone does this." Maybe. But how many countries form scientific organizations before they even have their first birthday as a nation? The Geological Survey of Canada formed in 1841, twenty-six years before Confederation! No wonder my parents, husband, siblings, children and grandchildren pick up rocks everywhere we go. Evidently, that's what Canadians do.

But before all that —before much of anything we know about yet — our universe was born. It happened about 13 billion years ago, when a wild burst of energy, beyond our imagining, exploded. Maybe that helps explain our fascination with fireworks? I don't know, but because we can't wrap our minds around it, we simply call it The Big Bang. It's easier that way.

After The Big Bang, seething cosmic dust and gases clung together in wild space and eventually coalesced into a hot molten swirl that became the Earth's core. Heavy chemical elements formed the core of this swirl; lighter elements attached themselves around this, and then a lumpy thin crust sealed it, more or less. It might be compared to an egg, with the core being a runny yolk, the mantle being a semi-runny egg white, and the crust being the shell.

Volatile gases found vents in the crust and expelled heat, lava and gas into the atmosphere. Things were very *busy* in those days it seems. The gases were partly made up of acidic water vapour, which eventually fell on the baby planet as rain.

This rain, after a million years or so, created a watery planet. At the time that the Acasta gneiss was formed just over 4 billion years ago, the air had little oxygen and there was no ozone layer to provide a shield from the sun. It would have been a poisonous place. Scientists say that there is no evidence of life on Earth for the first half billion years of its existence.

For the first 500 million or so years, earth-quakes, buckling and the movements of continents caused that gneiss to slide beneath and rise above the crust periodically. It was as if the Earth was restless in those years; but then, all birthing processes can be seen that way perhaps. It was for me, giving birth to my daughter; it was for my daughter, giving birth to her son. As a farm child I had witnessed numerous cattle, pigs, cats and dogs giving birth, too. The birthing process is basically the same for all of us. We quake, buckle, move, push, shove, rest, move again, often with our own personal sound effects. Labour is powerful, painful, intense, messy and miraculous. From that standpoint, I have a lot of empathy with Earth.

If you look at a map of Canada, above the western provinces, you'll locate the Northwest Territories (NWT), between the Yukon on the west and Nunavut on the east. Geologists further distinguish the NWT by naming different geological provinces within territories. Because there are no highways, cities, towns or even villages around it to use as landmarks, to find the Acasta River, look in the western part of the Slave Geologic Province, just south and east of the largest lake entirely within Canada, Great Bear Lake. Look specifically at 65'10"30'N - 115'30"30'W. That's where you'll find the Acasta River gneiss at this moment in time. It hasn't always been in this location, as you will see in Chapter Five.

At some points, according to Yellowknife prospector Walt Humphries, what we see of that Acasta gneiss outcrop today would have undoubtedly been beneath a mountain. Eventually, though, it was exposed and sat for a time bare and quiet. There was no willow, spruce, pine, nor the lichens, mosses or cranberries that live there today. The rock was bereft of wolves, caribou, bears, foxes, eagles, ravens and owls that dwell there now. Neither did the fat trout leaping in the river today have ancestors back then.

Today, almost one hundred bird species migrate to the Arctic to breed; many rest or breed in the Acasta River area, so in addition to the hum of insects, the land here sings a wild symphony. It wasn't always so. It must have been rather lonely in those first eons, although volcanoes aren't quiet, and the Aurora Borealis and streaking comets would have provided a light show, at least.

In *Shield Country: the life and times of the oldest piece of the planet,* Yellowknife naturalist Jamie Bastedo writes that, "The Slave Province is the oldest of the northern geological provinces, the most intensively studied. Internationally acclaimed as a showcase of Precambrian geology, the Slave, like all other shield provinces is made up of three kinds of rock: igneous, sedimentary and metamorphic."

If you haven't thought much about rocks since high school, here's a refresher. When magma is slowly pushed to the Earth's surface, or shot up by volcanic action, you have igneous (meaning *fire*) rocks. Obsidian and granite are igneous rocks.

Sedimentary rocks are formed chemically or physically from bits of rock, shells, fecal matter and plants that settle to the bottom of oceans, rivers, lakes. Water pressure and millions of years make the pieces solidify. You can often find fossils in them. They may also be formed and sculpted by the wind, such as you find in the Arizona desert or Banff National Park. Limestone and sandstone are examples of sedimentary rock.

Metamorphic rocks are those that began as sedimentary or igneous rock and then changed into something else. For example, limestone can turn into marble under the right pressure and heat conditions. The Acasta gneiss is metamorphic.

The Canadian Shield is the largest geological area of Canada, covering parts of the Yukon, NWT, Nunavut, Manitoba, Ontario, Quebec, Newfoundland and Labrador. To picture the Shield, think about the paintings of the Group of Seven and Tom Thomson. Those artists would have maybe painted fishermen and ballerinas if they hadn't fallen in love with the Shield. The Group broke new artistic ground in and for Canada and helped us see the beauty of our rugged landscape with fresh eyes. Until they and Emily Carr came along, much of our art had been along the European styles. Their interpretations of our forests and the Shield changed all that.

"The Acasta gneiss..." writes Bastedo, "is the nearest thing anyone can find of the earth's original crust. This in itself is exciting. But more than that, it holds the key to understanding what was going on in the earliest days ...."

Bastedo says that "in its subtle textures and skewed layers, the gneiss is a 'crystalline time capsule.' The Acasta gneiss could be to Slave geologists what the Rosetta stone was to early interpreters of Egyptian hieroglyphics."

It's worth noting that when Bastedo published *Shield Country* in 1994, it was believed that the Acasta gneiss was 3.96 billion years old. According to Wouter Bleeker, research scientist with the Geological Survey of Canada, new technology has now dated the rock at 4.03 billion years. Because we are constantly pushing the limits of science, it is likely best to say that the Acasta gneiss is the oldest *known* rock to date. When I spoke with geologist Tom Feuchtwanger in Calgary, he said that while he was in university in 1974, the oldest known rock on Earth was in South Africa, dated at 2.7 billion years old. And that's not ancient history. Tom is not, I mean!

It seems funny to me to worry about that little amount of time over the eons, but it's a big deal in the geological world. And, of course, everyone likes to know what is the biggest, fattest, longest, deepest thing on the planet. Otherwise, the *Guinness Book of World Records* wouldn't sell as well as it does. (And yes, the Acasta River gneiss was listed there by Jack Walker, the man who owns a claim on the rock. You will meet Jack Walker in Chapter Four. But for now, we return to our geology lesson.)

The first geologic eon is called the Hadean, which was when all hell was breaking loose with volcanoes erupting and lava spewing and meteorites blasting the planet and all that. Hadean actually translates as *hell*. The Acasta gneiss is from this time. Next came the Archean Eon when bacterial life began. And after that came the Proterazoic Eon, notable partly because stromatolites made the scene. I quite love stromatolites. They are very beautiful, and as you will see, they are responsible for making sex possible.

Stromatolites are huge floating algae beds. They began putting oxygen into the atmosphere through photosynthesis and began the creation of the ozone layer. The ozone layer helps shield our planet from direct sun rays, otherwise we'd be fried. You could say that life was made possible on Earth because of algae breath.

With the formation of the ozone layer, life on the new planet had a chance. Now that it's thinning, we all have to run around with hats and sun block. I learned from Tom Feuchtwanger that stromatolite beds still exist today and it occurred to me that they could be Nature's insurance policy. If we really mess things up on the planet and life has to start over, it will have a head start with these algae beds replenishing oxygen and making all things possible again. There, now. Do you see why I love them?

The Proterazoic Eon began about 2.5 billion years ago and ended about 500 million years ago. I'd say that's a rather long time to be pregnant with new life. It was during this eon that the Precambrian Shield more or less solidified. The Slave province smashed into the Churchill province next to it, and they've stuck together since. "As the new continent was being born," writes Bastedo, "the Shield's creation story ended. Here, the forces of destruction — weathering and erosion — have had the upper hand ever since."

Although the Shield solidified, that didn't mean it was — or is — stationary. In various sizes, shapes and configurations the continents have been moving around the globe since the beginning of time. Geologists have worked on explaining the tectonic plates and continental drift only for the last century. (A snap of the fingers in geological terms, of course.)

Wouter Bleeker explained that while the tectonic plates have been moving, they're also breaking up and colliding with other plates. "Think of a bathroom tile," he said. "You drop and break it, the pieces scatter. You may glue some back together again, you make a new tile out of the old one. But you drop it again, it breaks and you glue it again. After a while it becomes very hard to know where the original pieces came from."

These days, Wouter says, we have the Global Positioning System (GPS) to help us understand Earth better. Today, with satellite systems we can actually track the movements of North America. The records of the oldest cycles, of course, are fragmentary.

It was only about a hundred years ago that scientists began to explain how the continents have joined, drifted apart, rejoined, moved about, formed mountain ranges. In the 1950s using new information from magnetism, rocks and fossils, scientists were able to theorize that the continents had been one big super continent in the beginning. German astronomer Alfred Wegener named this *Pangaea.* This super continent, over millions of years shifted and drifted all over the map, as it were.

"Those particular rocks in the Acasta area," explains Wouter, "are about 4 billion years old and there are other rocks in India that have characteristics that are remarkably similar. We are getting to the stage now where, even for these oldest ones we are starting to have some indications about where some of the missing bits are. But how did they get there? There's still a long way to go to sort it out."

The movement of the continents over time intrigued me because of my special interest in that outcrop in the Northwest Territories. I wanted to keep my eye on it as I imagined the continents in their global synchronized swimming routine. Could the gneiss have travelled around the world? Wouter said it was possible. And, furthermore, it could have crossed the equator several times.

When I imagined that rock being at the equator, I instantly thought that it would have had a palm tree or two on it, maybe a little bougainvillaea. But my expert witness reminded me that there was no vegetation on the planet in those days. It would have basked in the sun without a bit of shade in sight.

Scientists tell us that North America is drifting westward now. But I don't expect to see my nieces who live in Asia any time soon. Continents move approximately as fast as your fingernails grow.

Geological time is mind bending. News outlets run constant stories about climate change and refer often to what has happened "since the last Ice Age, 10,000 years ago." The more I read about the history of our planet, the more I began to think of 10,000 years ago as yesterday. Also, I hadn't realized that there had been several Ice Ages; I must have missed the word "last" all my life.

I live only a few hours from the Columbia Ice Fields in the Rockies and have seen the retreat of the glacier there. It all looks rather orderly. Parks staff have even put little markers along the path so you can see where the glacier's toe was at various times in history. I can tell you that it's a much longer hike to the toe of the glacier than it was when I first saw it only forty years ago. It is disappearing fast. But it also looks like a deceptively neat, and so I hadn't realized that the Ice Age didn't end in a nice tidy retreat.

Bastedo writes of a period of 50,000 years in which the ice retreated and advanced repeatedly. "Full grown forests likely became established during the warmer spells," he wrote, "only to be overrun and crushed by surging lobes of ice as the climate cooled once again."

Fast forward from the algae floating around in water to prehistoric animals wandering about. The imagination reels, thinking about prehistoric animals living and dying right where we live today. When I visited Trafalgar Square in London it had never occurred to me that it had once been the ancient home of hippopotami and lions.

I do have a little clue about Alberta's prehistoric life because I live in dinosaur country in Alberta and enjoy visiting the Tyrrell Museum. But it's hard to imagine what prehistoric life could have traipsed across the Acasta gneiss because retreating glaciers scraped the area clean of remnants. Unlike the Alberta Badlands, the Shield is hard rock. Or perhaps we are not yet advanced enough to read the whole story in these particular rocks. It is likely, though, that woolly mammoths and sabre tooth tigers scratched their backs or rested in the shade of the Acasta River outcrop at some point.

"A little less than 20,000 years ago," Bastedo writes, "a mysterious force started pulling on the planet's climactic pendulum, triggering a gradual swing back into a period of warmth and renewed life." That mysterious force he mentions could be the force of life itself.

James Lovelock is a British inventor who developed many scientific instruments, some for the National Aeronautics and Space Administration (NASA). It was while working as a consultant for NASA that Lovelock developed the Gaia Hypothesis, for which he is best known. In his 1988 ground-breaking book, *The Ages of Gaia: A Biography of our Living Earth,* Lovelock wrote, "The idea that the earth is alive is probably as old as humankind. But the first public expression of it as a fact of science was by a Scottish scientist, James Hutton in 1785. (Hutton was a physician, geologist, naturalist and experimental farmer. It's hard to imagine that his theories were accepted very happily in those days; it was hard enough for Lovelock 200 years later.)

Lovelock, who named the study of the living Earth "Gaia geophysiology," taught that Earth is alive and that the evolution of organisms and of rocks need no longer be regarded as separate sciences.

The whirling stardust that became Earth and the Earth that became us is a story told by the ancient ones who had no scientific knowledge and also by naturalists, geologists and chemists who have much scientific knowledge. What has the gneiss to say about it all, I wondered. And then I found a story by Sleeping Crow which had an answer. Or, maybe not.

*Ala:Kalahanamae called me into his lodge late one afternoon. He looked serious, which worried me a bit. This was an Elder who almost always had a smile, or at least a smirk, on his face. "What have you learned today?" he asked, motioning me to sit.*

*I went on for a while, describing this, talking about that, chattering, really. I hadn't yet learned the true value of silence. He listened patiently, then suddenly began to shake his head from side to side. It was a clear sign to "shut up" and listen.*

*"Is a rock alive?" he asked, still very serious.*

*I thought for a few moments. The question seemed obvious at first, until I thought about it. Then, no answer seemed right. Where was he going with this?*

*"No, I don't think so. A rock doesn't move, doesn't hunt, has no eyes, and doesn't pray. I guess a rock is not alive, Father."*

*He shook his head from side to side once again, this time in disgust. He waved the back of his hand at me and sent me packing for the day.*

*A long time passed before Father brought up the subject again. During that time, I decided to find out what I could about rocks. It seemed silly since I thought I knew all about rocks. Still, there was much more than I ever expected. I even read physics manuals and researched elementary particles, down to String Theory. I was determined to be ready should Father ever ask that question again. Now, I really knew about rocks, or so I thought.*

*Father stopped me one day by the river. "Is a rock alive?" he asked, smiling now.*

*"Sure," I shot back. I launched into my newly-found "star stuff" argument, the principle of physics that tells us we are all made of the same elementary particles, which may, themselves, be alive — or even contain entire universes unto themselves. I was about to get to the heart of the matter, right down to String Theory, when he slapped me hard on the shoulder.*

*"You've learned nothing!" he shouted. "Get your ass out there for a week and talk to the rocks. Do nothing else but talk and listen. Just to the rocks. When you come back, come to my lodge and tell me if a rock is alive."*

*I was confused and, frankly, more than a little embarrassed. Father wasn't happy about rocks not being alive and he didn't seem to be very pleased about them being alive either. How was I going to get myself out of this situation?*

*I decided that the best place to talk to the rocks was somewhere in the mountains. I went to the Lassen area in Northern California, a land of honor to our ancestors. I camped for a week, alone, watching, listening and talking to the rocks. It was an amazing experience, and one that has stayed with me for this long life.*

*Rocks move! They change shape. They become round, disappear back to Mother, just like we do. They are created when Mother is angry. They are as varied as Mother herself. They are comfortable on the land, in the lakes, and at the bottoms of the streams and rivers. Rocks are everywhere! They dominate so much of our land, so much more than we can ever hope to understand. I had missed all of this because I had always moved too quickly through life. I had to learn to move at the speed of a rock to know the rock spirit.*

*I also learned the value of silence on this trip, and returned to my Father considerably older and a bit wiser. When I came back to our village, I went straight to his lodge with a gift of tobacco.*

*"Well," he mused. "Is a rock alive?" I smiled back at him, a copy of his own smile. "It's really a silly question, Father, isn't it? He nodded.*

Sleeping Crow isn't the only one to learn from rocks — or to feel a connection to them. All around the world there are people who connect to rocks. Or maybe it's the rocks that connect with people?

**Questions for Reflection**

1. What experience have you had with a rock, stone or geologic formation?
2. The Canadian national anthem contains the words, "The true North strong and free." What experience have you had with Northern Canada or Northern Canadians?
3. When has a rock or mountain has been alive for you?

**The View from Space**

"From space, I get the definite, but indescribable feeling that this, my maternal Planet, is somehow actually breathing — faintly sighing in her sleep— ever so slowly winking and wimpling in the benign light of the sun, while her muscle-like clouds writhe in their own metric tempo as veritable tissues of a thing alive."

Guy Murchie, pilot, author, artist

Chapter 3

# Connections

**[The stone] is a palm-sized oval beach cobble whose dark gray is cut by a band of white which runs around, and presumably, through it; such stones we call 'wishing stones,' for reasons obscure but not, I think, unimaginable.**

Annie Dillard in *Teaching a Stone to Talk*

Around the world people have long understood rock as a metaphor for the sacred. Stories, scriptures and hymns abound with this imagery. Followers of many spiritual traditions venerate particular rocks and stones. I wondered, What is this connection? I imagined inviting storytellers from everywhere into a Circle to share their tales of birthing stones, monoliths, healing stones and sacred mountains.

This chapter could be much longer; the numbers of stories seem endless. Undoubtedly, you will recall other mountains, formations, stones, rocks and sacred sites that you know about. I begin this overview with a story from the Northwest Territories.

### The Legend of Bear Rock

If you travel down the MacKenzie/Decho River, near where it meets the Bear River, you will see three large circular reddish brown images on a high cliff. This is known as Bear Rock. The marks resemble huge stretched beaver hides. In the time before history, three giant beavers emerged from Great Bear Lake and went wild. First they created huge rapids in Bear River, making it difficult for canoes to travel there. Next they began hunting down and killing the Dene.

The Creator took pity on the humans and sent to themYamoria, a great law giver. The Creator had in mind that Yamoria would bring order to the world and make things right between humans and animals. Yamoria faced the giants and killed them, one by one, near what is now the town of Tulita. He stretched the beaver hides on Bear Rock and left them there forever. He next built a fire to cook the animals. Some of the beaver fat dripped on the land. Today, smoke still rises near those spots. Legend has it that people who see fire in these smoky places, will live a long time.

The legend of Bear Rock is known by all five Dene tribes of the Northwest Territories. As stories do, it helps to unite people.

**The Holy Stone of Siberia**

In his intriguing book, *The Raven's Gift,* adventurer and research chemist Jon Turk recounts his journey to The Holy Stone of Siberia. The Holy Stone, he was told by an Aboriginal elder, "is the centre of our power and our magic." Turk, like many western scientists, was skeptical. But not skeptical enough to refuse the difficult journey, nor to ignore the instruction to leave a gift for the stone. Nor was he skeptical enough to dismiss stories of the stone's history.

A medicine woman, Moolynaut, told him that the Holy Stone held power, both for individuals and for the community. "Even enemies came to the spring festival [at the Stone] because this was a time of peace," she explained. "No one was allowed to fight at the Holy Stone. Sometimes when the festival was over warriors went away and then became enemies again. But sometimes enemy warriors decided to marry a woman from another clan or they decided that it was too much work to be enemies anymore, so they became friends. Things happened that way."

Turk's journey from North America to Siberia changed his life. He had come to be healed of chronic pain and to learn from the people of this land. He was shown a world unknown to him, but almost recognized. Part of the power of the Stone was explained to him: "...the mayor told me that energy flows from the magma-filled bowels of the earth to the Holy Stone. At the same time, people absorb positive energy from the tundra and also feed it to the Stone. Thus the Stone is a focal point, a storage unit and a transfer station for good energy."

Turk was healed by the medicine woman through his journey to the Holy Stone. Back in the US, doctors confirmed his healing but could not explain it. Turk tried to logically figure it all out. His experience with the land and the people of Siberia confounded his western mind and scientific learning. In the end, he realized that we cannot know everything, or explain everything. He surrendered, and decided to rest in the mystery. Later, when he was thrown into deep grief after his wife was accidentally killed, he again went to the wilderness for healing. He writes, "A bear and a raven. Magic moments integrated with the greatest sadness, preaching acceptance ... teaching me to heal by finding wonder within tragedy." After reading his book, I concluded that a rock can lead you anywhere.

**The Manitou Stone in Alberta**

A few years ago I was told by a Cree friend about a holy stone in Alberta, my home province. It is The Manitou Stone. Learning about it made me cringe. The story is wrapped inside the larger painful story of European Christians believing that they needed to "save" people, and the establishment of Indian Residential Schools by the Canadian government.

In the early days, the story tells us, the Great Spirit Manitou, sent an iron stone to Earth. Some call it a meteorite; others call it Old Man Buffalo, perhaps because of its power to protect the buffalo, the life blood of the people of the Prairies.

The Blackfoot and Cree learned from Elders that the stone offered protection against war, disease and famine. People left offerings near the stone and made prayers and songs in its honour.

When Christian missionaries arrived in the area, they saw how the people regularly travelled to the stone and revered it. The newcomers believed that the Manitou Stone was a stumbling block to the conversion of people to Christianity. In the 1860s, Reverend George McDougall, a Methodist, spirited the stone away and left the land and the people, bereft.

The prophecy of the Elders came true. Very soon, there was indeed war between Blackfoot and Cree peoples. Within a few years, the buffalo population had been decimated.

And, finally, a smallpox epidemic swept the area, killing many, including George McDougall's three daughters. War, disease and famine had arrived with a vengeance.

The Manitou Stone had been sent east to Ontario where it was lodged at Victoria University for more than 100 years. Eventually, a new generation of church people felt shame for the theft and returned the stone to the West. But the councils of Elders and chiefs could not decide who should take responsibility for its care or where it should be placed. And so, the Manitou Stone today waits. It is on display at the Royal Alberta Museum in Edmonton.

Just before the Solstice in December 2012, a Ceremony of Forgiveness was held at McDougall United Church in Edmonton. Initiated by Anna Faulds of Cold Lake First Nation, the ritual involved representatives of Blackfoot, Cree and Dene nations, the Royal Alberta Museum and the local and national United Church. The intent of the ceremony was to remove the negative energy surrounding the Manitou Stone. Such is the power of sacred rocks to this day.

**Glacial Erratics on the Prairies**

There are many glacial erratics on the Prairies. These are gigantic rocks that were carried many kilometers by advancing ice during Ice Ages. When the ice melted, the erratics were left behind. Today they stand like a lonely surprise in a field, pasture or meadow, far from their original homes. The Okotoks Erratic just south of Calgary, for example, was carried from near the present-day town of Jasper in the Rockies, a distance of about 400 kilometres. Today it is a tourist attraction and a sacred site. One erratic near Elbow, Saskatchewan, the legendary Mistaseni, did not fair as well.

In the 1960s, the government set engineers the task of damming the South Saskatchewan and Qu'Appelle Rivers to form a reservoir which they named Lake Diefenbaker. The Mistaseni, sacred to the Cree, Saulteaux, Blackfoot and Assiniboia Nations, was located in what would become the bottom of the lake. At first, engineers considered moving the 400 ton erratic, but decided that it would be impossible. What they did instead, was blow it up. I imagine that it must have been akin to blowing up an ancient church or temple. According to an archeologist at the Saskatchewan Archeological Society, a cairn commemorating Mistaseni was placed near the lakeshore and pieces of it were distributed among interested individuals and groups.

**England**

The British Isles are rich with rock and stone stories, rituals and holy places. Stonehenge in England is by far the most well known of the megalithic sites. It can be seen for about three kilometers all around. There are about 80 major stones, some brought great distances from the area now called Wales. Although it is a very busy tourist and pilgrimage site, modern Druids are still able to practice ceremonies during Solstice. Stonehenge was apparently constructed so that humans could align themselves with the great powers of the universe. Stonehenge is approximately 5,000 years old.

**Scotland**

Scotland's Stone of Scone is also called the Stone of Destiny. In 1296, it became a war trophy, stolen by the English in an effort to break the spirit of the Scots. The Stone was taken to Westminster Abbey. It was an act clearly meant to humiliate the Scots, for the stone was a holy thing, kept at the Abbey of Scone where Scottish kings were crowned for centuries. The English renamed it The Coronation Stone and secured it beneath the throne upon which English kings and queens sat for their coronation. There it stayed undisturbed for years. That is, until Christmas Day, 1950.

Four university students believed that Scotland needed a token, a symbol of Scottish pride to revive their flagging nationalism. They decided that recovering the Stone would be worth risking imprisonment. Ian Hamilton, Gavin Vernon, Kay Matheson and Alan Stuart broke into Westminster Abbey, and spirited the Stone back to Scotland. They were eventually caught by police and the Stone was returned to Westminster Abbey. For a while.

In 1996, England finally returned it to its Scottish owners. The story underlines the authority and passion that people can invest in 150 kilos of rock. The story gained popularity through the 2008 British-Canadian film, *The Stone of Destiny.*

**Ireland**

Few visitors to Ireland wouldn't know about the Blarney Stone. This stone, in Blarney Castle near Cork, has many legends attached to it. One says that the Blarney was once part of the Stone of Scone in Scotland. Another says that the builder of a castle appealed to the goddess Cliodhna for help in settling a lawsuit. She told him that if he kissed a stone *enroute* to court he would receive an eloquent tongue. He did, and the judge ruled in his favour. In gratitude, he incorporated the stone into the castle. Today, the Blarney Stone is said to impart the gift of the blarney to the one who will learn over backward to kiss it. The literature proclaims that if you kiss the Blarney Stone, "you'll never be at a loss for words again."

**Easter Island**

The giant stones on Easter Island in the south Pacific have a very different story. These 900 carvings date from around 1500 CE. They weigh roughly 14 tons each and stand an average of 4 metres high. These imposing statues, *moari*, are mysterious. The work of carving them apparently ended abruptly but people have had difficulty piecing together their story.

The Dutch were the first Europeans to visit Easter Island, in 1722. It is hard to imagine what the sailors thought approaching an island ringed with monumental *moari*, their bright coral eyes staring out of elongated, giant faces.

Jo Anne Van Tilburg, archaeologist and director of the Rock Art Archive at UCLA, specializes in Polynesian studies. Working with the Easter Island community, they have concluded that moari must be representations of powerful men who perhaps acted as conduits for communication with the gods, like priests. Their great height, like the soaring cathedrals of Europe, may have been a symbol for the connection between heaven and Earth. Even today, the moari inspire awe. As does the Black Stone in Mecca.

**Saudi Arabia**

Tradition says that the Black Stone fell from the sky. It is said to have been venerated in pre-Islamic times, and according to Muslim tradition, dates back to the time of Adam and Eve, the beginning. It is the eastern cornerstone in the Grand Mosque in Mecca, Saudi

Arabia, set there by the Prophet Muhammad in 605 CE. A piece of it is framed and set into the side of the mosque and this is where millions of pilgrims annually come to pray, and to kiss, touch or point to it in a show of reverence. Pilgrims walk around it in prayer seven times. If devout Muslims can possibly do it, they are required to make at least one pilgrimage during their lifetimes. Many millions do just that; there is but one Black Stone. For Buddhists, though, there are many Mani stones.

**Asia**

Buddhists in Tibet, Nepal and India set Mani prayer stones in piles, along river banks and on walls. These are stones, pebbles or rocks that may be decorated with pictures or words, or left in their natural state. Most are inscribed with the traditional prayer for compassion *Om mani padme hum*. The stones offer an invitation to anyone passing by to pause, to pray and to add another stone or rock. When leaving the site, people walk to the left side, clockwise, as a mark of respect.

**USA**

Throughout North, South and Central America there are stone circles, spirals and buildings that honour Earth, mark the seasons and align with planets and constellations. According to author James A. Swan, the Serpent Mound in Ohio is "one of the most spectacular shrines to the Earth god/goddess in the world." Created about 2,000 years ago, the form stretches and coils almost 400 metres. Built from river stones and packed with clay, the snake is the largest Earth sculpture in the world. The ancestors of the Snake Clan of the Hopi Nation built it. They would be amazed, perhaps, to know that today a three-story observation tower has been built so that visitors can view the whole sculpture at once. In this case, it is the stones that have been used in a reverential manner. The pipestone in Minnesota is itself considered sacred.

The pipestone is catlinite stone. It is reddish brown, easily carved. It is inaccessible except by quarrying down through quartzite. For centuries, this pipestone has been carved into long stemmed medicine pipes for sacred ceremonies. Today, the quarries are protected areas because the stone is considered sacred.

**Jerusalem**

Some rocks in Jerusalem are considered sacred — and contentious. The most well known is the foundation stone, housed in the Dome of the Rock, a shrine on the Temple Mount. The Dome of the Rock was completed in 691 CE. Its dazzling shiny dome is prominent today in many of the photographs that tourists and pilgrims take home from Jerusalem. A story in the Muslim tradition tells that Mohammed flew from Mecca to this rock on what is known as his mystical Night Journey.

The site is also considered sacred in Judaism. Here, King Solomon built the first Temple in 957 BCE. This rock is where Isaac was brought for sacrifice, and where the Ark of the Covenant (housing the Ten Commandments) once stood. This stone in Jerusalem is a very public and storied one. In far off Canada, a writer and carver worked on a very private one.

**British Columbia**

Stone carver, poet and teacher Ross Laird, offers another glimpse of the power of rocks to engage our spirits. His book *A Stone's Throw: The Enduring Nature of Myth* chronicles his love affair with a rock, and much more.

With his father, Laird climbs a mountain searching for a rock to carve. Over a period of one exact year, he works on his rock in his studio. As he carves, he also writes, inviting readers along on his mythic, spiritual and personal journeys. At the end of the year, Laird returned the carved stone to the wild. He concludes, "This stone that I have worked, that beckoned in dreams, will be buried in waters of black fire. It will find solace in the waters. It will be lost, and found again by its homecoming. 'Stones,' said Pythagoras, 'are frozen music.' Perhaps it will sing, and I will hear its song, far off, carried on the wind."

**Australia**

Uluru, once called Ayers Rock, is a sacred site, a tourist destination, a wonder of the world, a geological beauty and more. For forty thousand years it has figured in Aboriginal ceremonies, of their understanding of the world and of Dream Time. This enormous sandstone rock provides physical as well as spiritual life for the people who hold it sacred. Author and photographer Courtney Milne writes in *The Sacred Earth* that Uluru "has become an important earth symbol with newly emerging mythologies, and it is visited by increasing numbers of non-Aborigines each year. Its power, however, lies in the unbroken chain of Aboriginal worshippers stretching back into the mists of time."

**Hawaii**

I chanced to read an article by Hazel Leung who had gone to see ancient birthing stones in Hawaii. She had learned that royalty came here to give birth. I was struck dumb, imagining giving birth with birds singing in the trees, a gentle breeze caressing my hot forehead. (It wouldn't be raining, of course.)

Leung wrote that people still care for these rocks in the manner of natural sacred sites and that flower leis had been draped on them. These Kukaniloko Birthing Stones are on the Island of Oahu.

The stones are in the middle of a rural area. "There are no special fences or signposts, a simple sign tells you its name and that it is a sacred site." She reported feeling a sense of peace and of seeing the profile of a pregnant woman in the distant mountains.

"Then I was suddenly hit with a rush of energy," she wrote, "like goosebumps only stronger, it rushed up my legs and filled the area of my abdomen... I felt light and happy... as if I was grounded to the earth."

**Tanzania**

There are many mountains around the world that are considered sacred. The highest one in Africa is one such mountain. Mount Kilamanjaro is in fact, three volcanoes — one collapsed and two dormant. Rising 5,895 metres above sea level, it was named by the Masai, *Ngaje Ngai*, meaning "House of God." It is visible for hundreds of kilometres in every direction.

It was in the shadow of this sacred mountain that our human ancestors first stood on their hind legs and walked. I imagine that the awe that the massive peaks inspire today was felt in the hearts of our distant relatives, too. Mount Kilamanjaro today is not only a tourist destination, but is considered a sacred pilgrimmage site, as well.

**Personal Connections**

During the research for this book, I asked my husband Bill Phipps what he thought about our human attraction to rocks. He pulled out the red and black stones he always carries in his pockets. One is from Haida Gwaii, an island off the north coast of British Columbia, and the other is from the shores of Lake Superior. Some would call them worry stones, likely, but Bill calls them 'reminders.' They remind him of the sacredness of Earth.

When he celebrated his seventieth birthday recently, he offered reminders to others. After the candles and cake, he placed fossils he'd collected 30 years ago from Lake Superior on the table and invited people to choose one. He spoke of being grounded, of linking past and future. "These connect us to our Earth home," he said.

I asked many people about our connection to rocks generally and to the Acasta gneiss specifically. Some scoffed at the idea of any stone being "sacred."

A geologist cut off the question, and declared he was a "very down to earth person"! A stone cutter merely smiled when I asked if he had any special feeling about the rock. An Elder said, "Yes, of course; all Earth is sacred. And all Earth has energy, too."

I wondered about rocks having energy. Jon Turk is one who became a convert to that idea and there are millions of others who believe in the power of rocks to heal. A visit to a local rock, gem or health food store attests to this. There, we can learn about Mongolian Holy stones, or the soothing effect of quartz, or the stimulating effect of magnetic hematite, and much more. I was curious enough to ask Reiki master and healer, Mary Botel what she thought.

**Reiki Master Mary Botel**

Mary and her sister Sandra (also a healer) grew up on the west coast of Vancouver Island. They each worked with a piece of the Acasta gneiss and came to the same conclusions about its energy. They stated that they felt it was a powerful rock, with healing properties. I interviewed Mary in the spring of 2012 after she had had a piece of the Acasta gneiss in her possession for about a year.

"The first time I picked it up, it was buzzing with energy," she said. "The first message I got from it was that it was about the connectedness of all, about remembering the interrelationship of everything. If we don't help each other we aren't going to survive the mess the world is in."

As she held the rock, open to its messages and vibrations she said that she received "vast brain pictures of other planets, the universe... I could see green, purple, whitish beings, light beings smiling and open. This happened to my sister, too."

Mary believes that these light beings are healing entities. They want people to understand their message that we are all interconnected, "that there's been a disconnect with the people and with the natural planet rhythms. That's why things are unwell now. People are losing their way. They want us to understand that nobody owns the land. The land is a gift for people to use wisely.

"We need to remember that love and compassion are important. There needs to be a reconnect with Earth's natural rhythms; these frequencies need to be realized. Earth vibrates at a certain frequency."

Mary uses crystals in her healing, meditation, energy work, and as an aid to concentration. Rocks have energy that works in different ways. For example, "if I pick up a rose quartz it has a lower, slower vibrational level, a singular energy. An amethyst is a common crystal that people use to to help clear the brow chakra, it's got a clear, lovely feel to it, an even energy tone, it can help people receive visions, it can amplify psychic abilities, it helps to tune in to other people." I asked if the gneiss was similar to other rocks she has studied. "No," she said. "But this rock is filling me with other information; it's very busy.

"It has healing energy that is not specific. The healing energy of crystals is specific; the gneiss has a wide spectrum of healing power."

Mary's understanding of the world reminded me of how Jon Turk explained the world to himself when he was trying to figure out how and what he believed: did he come from an understanding of magic or of logic? People who receive information directly from Nature have a different perceptual *gestalt* from those who receive it primarily from books and computers.

Mary received her early learning directly from Nature. Her home community of Winter Harbour, on Vancouver Island, had no road, electricity, radio or television. "I listened to the humming of the trees. I walked the beach and gathered rocks that needed to speak. My *self* didn't have to block out any extraneous energies. I didn't know that everyone else was not like that." It's difficult for her in the city now with all the "noisy" electricity. She calls it sensory overload.

"Kids today have to learn to block it out but then they also end up blocking out the natural energies."

The main message that Mary relays from her reading of the Acasta gneiss is that we are all connected. Cosmologist Brian Swimme tells us the same thing, and reminds us that ancient stories gave us this information long ago.

As I write this, Canadian astronaut Chris Hadfield, commander of the International Space Station, is circling Earth at 28,000 kilometres per hour. He has shared his view of our connectedness dramatically using all the social media available, by speaking directly with students across the country and by recording "Is Somebody Singing" with the rock group Bare Naked Ladies. One of his lyrics was, "You can't make out borders from up here." If rocks could talk, would they say the same thing?

For many of us, a connection to rocks is like an age-old love story. The Acasta gneiss, so recently discovered, adds another chapter to the story and provides a new lens through which to view our ongoing quest to know ourselves.

**Questions for Reflection**

1. What do you make of rocks, stones, gems or even mountains having healing energies?
2. What do you think inspires people to become rock collectors?
3. What other special rocks or stones or places do you know about that were not included in this chapter?

Arctic Wolf

Chapter 4

# On the Rock: Who Goes there?

**It is there on this littoral that the little Geological Survey of Canada aluminum silo sits — with spiked, bear-repelling covers over its port holes and a hand-painted wooden sign over the door: *Acasta City Hall Founded 4.0 Ga***

Daniel Wood in *Way Out There*

Northern Canada is often viewed as a treasure chest of riches. Oil, gas, diamonds, uranium, gold, silver and copper have enticed people to cross the 60th parallel for generations. Furs, fish, whales and Northern Lights have lured people for centuries. Adventurers have been travelling North since their ships could make it across the sea from Europe. Once the airplane was invented, there was no holding back. These days, international tourism draws people for a myriad of reasons.

The honeymoon of Prince William and Princess Kate is but one example of moderns seeking something a little out of the ordinary. Planeloads of tourists, including many Japanese couples hoping to conceive a baby under the Northern Lights, regularly land at the little Yellowknife airport. Of course for tens of thousands of years, for Dene, and Inuit, the North was and is, simply *home.*

One great privilege I had in Yellowknife was transcribing Dene Elder George Blondin's stories. George, born in the bush in 1922, was a well respected leader from the Sahtú region, around Great Bear Lake. His daughter brought his work to my home in Yellowknife in the early 1980s. "Can these words become a book?" she asked, handing me a box of scribblers. The scribblers were filled with legends, teachings, medicine stories and memoir, all carefully written in pen and pencil. I gained an education reading them. The scribblers became a manuscript and when more stories were added, it eventually became the book, *When the World was New: Stories of the Sahtú Dene.*

The Blondin family lived and worked in the general area of Sahtú, Great Bear Lake, the lake into which the Acasta River indirectly, and eventually drains. The oldest rock is in Tlicho territory south and east of the Sahtú, but I include this story to give a glimpse of the general landscape.

"Pagot'ine's Wound" is one story from George's book that gives a flavour of life on the land, and includes George's own birth. Karkeye and Paul are George's grandfathers. I was struck by the numbers of animals George mentions. Since 2010, there has been a hunting ban to protect dwindling caribou populations.

*When everybody left Tulit'a for the summer, Edward Blondin and his wife Julia decided to go to the far side of Sahtú, close to the Barren Lands. Edward's father, Paul, went with his wife to the head of Chukezedeh, to hunt and trap.*

*Edward camped in the same place as his father-in-law, Karkeye. He and his wife put up fish for the winter, and when winter came, they trapped marten and fox in the bush.*

*After Christmas, people who had heard there were many caribou on the Barrens started to come from all over — from K'ahbamitue, Déline, and the northeast shore of Sahtú. The people travelled to a place where there were many white foxes in addition to caribou. In the forests of a deep river valley, they had drum dances and hand games before going out to hunt caribou and trap white foxes. On one of these hunting trips, there was an accident.*

*A middle-aged man from K'ahbamitue named Pagot'ine was chasing caribou with a loaded gun when he slipped and fell. The gun went off, causing a wound in the top of his head that was so big the people could see his brain. All they could do was tie rags around his head, lay him in a toboggan and haul him home.*

*Pagot'ine was unconscious for two weeks, and people had to stay with him all the time. They took turns. Late one night, Edward Blondin was sitting with him when Pagot'ine woke up.*

*Edward saw that the sick man had opened his eyes, so he spoke quietly to him. "Uncle, it's me — Edward," he said. "Do you know me?"*

*The man's eyes seemed to flicker in answer, so Edward told him what had happened, explaining that it had been two weeks since he had shot himself.*

*Pagot'ine tried to speak. He told Edward that all he had heard was a great noise. "But I did have a second to think. I thought about Sahtú as a man in my medicine before I was born. I saw him: he was old. That was all I had time to think about until now.*

*Pagot'ine went on to say that he didn't think he would die. "If I die, bury me and laugh at your uncle because he lies!"*

*From then on, he got better and started to talk to people again. He lived until he was 96 years old, when he died of natural sickness and old age.*

*When Pagot'ine recovered from his wound, he told the people not to let his accident spoil their fun. The trappers and hunters had a drum dance and hand games. When it got warm, people went back to their own country — some to K'ahbamitue, some to Sahtú.*

*Because of the accident, the people hadn't stayed long in the Barren Lands, but Edward Blondin managed to trap 165 white foxes. Karkeye and his daughters trapped 275.*

*The Dene travelled at night when they went back to Sahtú in May. The snow was wet and slushy during the day, and they could move better when it was frozen. On this trip, Edward and his wife stopped for two days for the birth of their son, George Blondin. Then they kept right on travelling, as if nothing had happened.*

*When they reached Sahtú, the people hunted beaver and muskrat while they waited for the ice to melt. In winter, the lake ice becomes two metres thick in places, so that it doesn't melt away until July. After the ice finally cleared, the families set out for Déline.*

*The distance from the north shore to Déline is about 300 kilometres. Paddling that distance in a big canoe takes many days, so whenever the wind was behind them, the voyagers used the sail. They didn't stop to sleep because they wanted to get home as quickly as possible after being away a whole year.*

*At Déline, Edward Blondin learned that his mother had been sick since Christmas. He set out right away, heading down Sahtú De to Tulit'a, and he was able to see his mother just before she died.*

*As she lay in her tent, his mother told Edward that a woman in white had come to her in a dream. This woman told her she was a good person who had never complained in her life. Because of this, the woman in white had granted her three wishes. Edward's mother told the dream-woman that she had lived a long life and didn't need to live any longer, except to see her son and pass on to him some of her powers. So Edward had arrived.*

*His dying mother said to him, "If you have problems in your lifetime, I will come to you at least three times." She continued speaking to the people all day until she died that night.*

*The next time they went trapping, Edward's people travelled to Turilideh, about 160 kilometres from Déline. Whitefish swam downriver in fall, and the people were able to catch their winter's fish supply by putting a trap across the river.*

*Before freeze-up, they set up their winter camp, shot moose while they were still fat, put up fish, and cut winter supplies of firewood. Paul and Edward Blondin usually got up at five in the morning and worked all day until dark.*

*The families enjoyed living on the land, going where there was game, always moving. Sometimes, early in the morning, an old man or woman would sing Dene love songs to make the younger people happy. When the people travelled in the bush with their dogs, they sometimes came to a hill so high they could see the land all around. There, they sang a Dene love song to express their thanks to the Great Spirit. They were happy in those days.*

For George Blondin's family and generations of Dene, Métis and Inuit, the North is home. But the North has also been romanticized from the beginning of visits from others. When European explorers such as Martin Frobisher, Alexander MacKenzie, John Franklin, George Back and Samuel Hearne arrived nearly 300 years ago with their sketchbooks, mapmaking equipment and curiosity they began a era of legendary storytelling that has stretched through the years to Farley Mowat and beyond.

The Canadian government has always sent geologists north, including Joseph Tyrell of Alberta dinosaur fame. The Smithsonian Institute began sending their collectors north 150 years ago. Today, many visitors' names are known across the North, some for their folly, some for their cunning and some for their determination. But no matter how you tell the stories, the North has never been a place for the faint hearted or the sissies; fifty below zero is no picnic.

When gold was discovered in the Yukon and the 1898 rush was on, it prompted more activity in the Northwest Territories, too. Gold was discovered near the site of today's Yellowknife, but it was not the Yukon kind of gold; this was the hard rock mining kind. This meant that the gold and the Dene who lived around the north shore of Great Slave Lake, were basically left alone. As I wrote in my first book, *Yellowknife,* "Stories of gold at Yellowknife faded to mere rumours. The rumours remained whispers until the 1930s when technology caught up to the North. The drone of tiny aircraft was heard in the Northern skies as the work-horse of the North began its career. Planes brought mail, medicine and prospectors, many of them jobless men, desperate for a dream.

"Pitchblend, containing uranium was discovered at Great Bear Lake in 1931, following oil strikes at Norman Wells. Then gold was rediscovered at Yellowknife, and when its presence was confirmed by Dr. A.W. Joliffe of the Geological Survey of Canada in 1935, the news was accidentally leaked and ... *Boom!*

"Men left dustbowl farms and Depression-wracked cities, packed a year's supply of food, prospecting tools and headed for the North. By barge, scow and floatplane they came. Women followed. Few struck it rich, but hundreds of claims were staked and two mines, Negus and Con went into production. The first gold brick was poured at the Con in 1938. By 1940 work had started on seven mine sites."

Port Radium on Great Bear Lake yielded uranium, and Americans used it in their desperate bid to make the first ever atom bomb. In 1945, it was used to end Japan's participation in World War II.

Since that time, many more mines have opened (and closed) around the NWT. The greatest change on the mining scene next came in a huge staking rush in 1992 when northern diamonds were discovered near Lac de Gras. It was around this time that the world realized that a dreadful new term was being used to describe diamonds from Africa: *blood diamonds*. In contrast, Canada proclaimed that our diamonds were mined, cut, polished and set under fair and decent working conditions. In less than a decade, four diamond mines were in production.

"The true North strong and free," is about many things. It's a mythical place where poets such as Jim Green, musicians such as Leela Gilday, carvers such as Sonny MacDonald and artists such as

Goota Ashoona flourish. It's a harsh place where hard drinking, and more recently, hard drugs take their toll in high suicide rates, early deaths, high rates of addictions. It's a beautiful place of dancing Northern Lights, stunning shorelines and delicate beauty. It's an extreme place where the temperature can drop to 50 below zero in winter and where the summer midnight sun can bake you like a potato. It is adventure, modern land settlements, unique history, warm hospitality. The North is also about a government that honours diverse voices. In a population of 44,000 people, there are eleven official languages.

And, of course, the North is about rocks. It's a place that prospectors love. As science writer Bob MacDonald said, "the Shield is like a candy store for geologists" — and for other people.

Fred Sangris lives in Ndilo, Yellowknife and is a former chief of the Yellowknives Dene Band. He is a cultural historian who works with the Elders, collecting traditional knowledge. Like his grandfather David Sangris, he started life as a Barren Lands trapper in the Tlicho territory. His knowledge from his own experience and from the stories he's learned could make a set of encyclopedias — or several sets. Fred received a piece of the Acasta gneiss and talked rocks with Yellowknife journalist Aggie Brockman in June 2012.

Fred knows the land from many angles; one is by the geology of the areas he's travelled. From the Slave River to Great Bear Lake to the Coppermine River, he knows where the copper, diamonds, gold, soapstone can be found. He knows where to find the best rock for making knives to fillet fish, scrape a moose hide, build a cairn, carve a pipe or conduct a ceremony. He has great respect for rocks.

The Yellowknives received their name because they were known for the fine quality copper knives they crafted. Fred has a story about that.

"My grandfather would go to the Coppermine River, like so many Dene," he says. "There was a huge slab of copper there, it was bent right over, coming up out of the ground. It was green and dark. The people would take a little piece so they could work with it, making kettles, that sort of thing."

This was the same area where the great Chipewyan Chief Mattonabee took the English explorer Samuel Hearne on his quest for a copper mine in 1770. Fred says that the people took explorers round and round close by the copper, but didn't want to show it to them. Samuel Hearne reported to the Hudson's Bay Company that there wasn't much copper after all. Fred's story then jumps to 1910.

"The last group of Yellowknives to go up got a little, but the family said that the copper was receding back into the ground, that there was just some on the surface. The Elders explained that the copper was there for the Dene to use, but when they took too much and didn't share it, the copper disappeared back into the Earth."

Fred also offered cautionary words. "The Dene say that the rocks are alive, there's a spirit in there. Everything is alive, right? It's OK to pick up loose ones, but don't break them off. And we don't throw them around. We have high respect for rocks. When we find tent rings, we don't use them again. We believe that there are spirits attached to them. Sometimes there are bad spirits, too.

"When we have ceremonies we invite the rocks into our tents. We put them around our camp fires. They're old. They know a lot of stories. Fred laughs and adds, "They might like to hear stories from us, too, eh?!"

The Acasta gneiss has certainly generated many stories since its discovery by a young Ph.D. student from Queen's University.

I have a beautiful photograph of a happy woman holding Jowi Taylor's Six String Nation guitar. The photo was taken at a concert in Yellowknife. She is pointing to an embedded rock on the fret, a piece of the Acasta gneiss, which she discovered. Dr. Janet King, who is now Assistant Deputy Minister at Northern Affairs Organization in Ottawa, was then working with the Geological Survey of Canada.

In a phone interview, in June 2012, I told her that my research into the Acasta gneiss made me feel very tender toward Earth. She said, "I love to hear that. I can share with you that I feel that way all the time; many geologists do. I don't know if they call it tenderness, but a deep regard and understanding of the Earth and everything that it is now and everything it has been over its billions of years of history. To me, it's one magnificent story."

Janet explained that for several summers she had worked on mapping the Precambrian Shield north of Yellowknife, so she was familiar with ancient rocks between two and three billion years old. This one summer, she was much farther north. She was camped with a team of seven others on an island in the Acasta River. "We'd been examining the rock for some time. It's a very complex gneiss. At first we didn't understand how it fit into the geological story of the area.

"Around the Acasta River the rocks are part of an ancient mountain belt that would have looked rather like the Rocky Mountains do now. The team was interpreting what the nature of the Earth in the area had been about 1.8 billion years ago."

As she speaks, I'm picturing the eight of them, young, excited to be on the land, far from the southern cities where they studied. Every day they load up heavy back packs with their lunch, pick axe, collecting bags, bug spray, bear bangers and more. "It's remote in the Barren lands. It might be a gorgeous day but you always bring as much warm, dry clothing as possible because a snow storm can blow in, or rain... you have to bring emergency equipment so that you can stay overnight safe and warm. It's thirty or forty pounds over ten or fifteen miles."

I imagine them as they take a boat to the mainland and set out in pairs following an almost-trail which has been marked on a map and aerial photo. They walk most of the day, stopping to record observations and collect interesting specimens (thus adding to the weight of their packs. Oh, to be that strong and vibrant.)

It so happens that on this particular day, Janet was paired up with a man. "This gets a little corny," she laughs, "but I was with my husband-to-be. It was one of the few times we ever worked together, the two of us walking." I thought it was very romantic.

"You know that *Eureka* moment?" she asks. "It suddenly all fell into place." She is speaking not of the romance but of the rocks. Because of her earlier work with the old rocks near Yellowknife, she recognized that she was looking at something *very* old.

"Once the Eureka moment happened, we started exploring that old domain of rock, that really messy gneiss. [The team] characterized it as best we could and then sampled it. We took big buckets full of rock and sent it to a research partner Sam Bowring, our colleague in the United States. I went back to Queen's University and it became part of my thesis. It was pretty exciting when they contacted me to give me age of the rock.

"I was young, caught up in the world of science. I took personal satisfaction from having contributed in that way. When I think back, I had pure pleasure being out on the land with the opportunity to tell the story of the rocks. We created knowledge, and that created understanding. The work I did launched me into my next step as a scientist."

Dr. Wouter Bleeker, research scientist with the Geological Survey of Canada picks up the story for me. He explains that, "Once it was realized that these particular gneisses were very old, a geochronologist working on that project (an American Ph.D student) suddenly got rather interested and pursued their age dating in more detail. Still, to get an accurate age on these very old and much modified materials is not easy. It took the development of a very sophisticated piece of machinery to get close to the real age. (This was an ion probe configured specifically to date very small spots within very small crystals of zircons, which are a natural part of the rock.) This machine was being built in Australia at the time [of Janet King's discovery.]

"Finally, some of these very old samples went to Australia and resulted in an age of 3.96 billion years. At the Geological Survey of Canada in Ottawa, we installed our own ion probe in the early 1990s and spent more time on these rocks, refining the age to 4.03 billion years."

Dr. Sam Bowring, an American geophysicist from MIT and his teammates from Washington University of St. Louis, Missouri, the NWT Geological Division of the Department of Indian Affairs and Northern Development, and the Geological Survey of Canada all went to the site in 1985. (This research was also supported by NASA and the US National Science Foundation.) In 1989 the announcement was finally made. Yellowknife prospector Walt Humphries laughs telling a story about that announcement.

"When this great announcement of the age of the gneiss was made, Australians and Americans heard it first," he recalls. "The announcement was delayed in Canada because it hadn't yet been translated into French!" As a government announcement, this was imperative. But after 4 billion years going incognito, another few days were neither here nor there for the rock, I imagine.

Walt Humphries was still laughing, I imagine, when he and his partner geologist Brian Weir chartered a plane to the site. But it took some doing before their float plane lifted off Yellowknife Bay.

My husband Bill and I visited Walt at his home in Yellowknife in May, 2012. It is on Gitzel Street, just steps away from the Trans Canada Trail which runs along the shore of Frame Lake, and a few minutes walk from City Hall. Gulls, tern, wild canaries, ravens and robins sing spring from the birch, jack pine and spruce. The ice is off this small lake at last; everyone is celebrating it seems. Moms push strollers along the path, sharing space with cyclists, joggers, and people ambling along with take-away coffee cups.

We are early for the appointment with Walt, one of the most well known prospectors in the north, so we sit to enjoy the view for a while. This lake is a downtown treasure; the citizens have opted to build the stunning Legislative Assembly, The Prince of Wales Northern Heritage Centre, the Visitors' Welcome Centre, the hospital and City Hall all around it.

Walt is known not only for his love of geology, which he willingly shares with school children and the rest of us, but for his paintings, his work as a mining exploration consultant and with the fledgling Mining Heritage Society. He is a writer, too. In his long-standing newspaper column, *Tales from the Dump*, Walt offers his opinion on just about everything that's going on in the North. City councillors who value their jobs, it is said, read him regularly. He is a fierce proponent of the human right to recycle and scavenge, and he believes that the dump in Yellowknife could lead the world in right-thinking about composting, as well as reusing and recycling.

Walt's home reflects his personality: warm, interesting, full of rocks, petrified wood, fossilized things, including dinosaur dung, and beautiful (and funny) art. Everything has a story attached to it, for above all, Walt is a storyteller. We drink tea and settle in his comfortable living room, watched over by a friendly calico cat. We are here to learn about his connection with the Acasta gneiss.

"It's hard to keep secrets in the North," he begins. "News got out that the find [of the oldest rock] was in the NWT at an undisclosed location," he says. "The geologists were going back and forth from Yellowknife picking up samples. It sort of irked me a little bit that they had pieces, while no one else did. But I heard that it was on an unnamed place on the Acasta River. Well, there are only so many spots there that you can take a plane into. Eventually I showed someone a map. I asked, 'Is that the place?' And he said, 'Yup.' Then I knew exactly where it was. If I hadn't had the map, he wouldn't have told me."

Walt had worked on other claims with Yellowknife geologist Brian Weir and they decided to partner on this venture, too. "Once we got out there, we could see where they'd been banging around. Brian loaded up the plane with rock and I staked a claim on it."

Humphries and Weir harvested fallen pieces of rock on several charter flights to the site over three years. They believed that people would want a piece of such a unique rock. Although too many of the souvenirs Walt constructed by mounting a piece of the rock on a plaque are still in his basement today, many people are interested in the rock. Among them, Vancouver writer Daniel Wood. The way Wood describes his time at the site makes it sound like a pilgrimage.

In his article written for *Way Out There: The Best of Explore,* Wood describes his journey to the Acasta River. After learning that his brother was dying of leukemia, Wood went looking for a touchstone. What better than a rock that offered perspective on time and our own place in the universe?

He describes flying in from Yellowknife on a Beaver float plane. The first things he saw upon landing were recent grizzly bear tracks by the shore. He quickly made another discovery. His touchstone was protected by armies of black flies and mosquitos, which he describes as being like hordes of "petulant Greek Furies."

Wood describes the landscape. "A dozen islets whose jack pines are silhouetted against the low horizon. The arctic sky is as pale as tears. It is quintessential Canadian Shield vista: sparse and elemental. If Tom Thomson had painted the scene, the tree branches would be outlined in crimson. Ahead, the island's curving beach ends in a 130 foot bluff. Large glacial erratics, some the size of refrigerators, litter the highest reaches of this ridge. More boulders litter the shoreline below the cliff. To my left ... a tangle of dwarf willow and blueberry."

Wood climbs onto the rock, and there on the unnamed island on the Acasta River he gathers samples, and gathers his thoughts about the eternity that has a claim on his brother.

When Walt Humphries and Brian Weir staked that first claim, they banked on the fact that most people couldn't afford the flight to Yellowknife plus the charter to the remote river that Wood made; they thought that the business of selling what they could harvest was logical. "Brian wasn't really comfortable with selling rocks," Walt said. "He thought that was more for a hobbyist; there was a schism there. But I marketed to schools, universities and museums. I felt that is part of the business, and students need them to learn.

"There are schisms, too, among mining people, geologists and rock collectors," Walt continues, warming to a pet peeve. "There was a mine at Pine Point [NWT] and they had world class, museum-quality samples, but they just threw them into the mill. It would break your heart if you were into rocks and minerals. I've seen museum pieces that were worth a thousand times more than the twenty cents of lead or zinc they got out of it. But most mining companies in Canada do not get into the rock and mineral trade. It's like pulling hen's teeth to get a geological sample out of them."

Walt's energy over this point has disturbed the cat, but we are all ears. He's right, I think; I'd just never thought of it before. He goes on to talk about the Giant and Con gold mines in Yellowknife, recently closed, and the Diavik diamond mine further north which has recently opened.

"Every mine in Canada comes across some really neat stuff. Where's the Con mine sample? Where's the Shield sample? It always drove me nuts about the mining companies. Where's the big chunk of kimberlite sitting in town for everyone to go to look at and learn about?" We pour more tea, and Walt sits back to receive another question. I ask him how long he held the claim at the Acasta River.

"After three years, we decided it was time to move on," he says. I wonder if I sense a tinge of regret in his answer.

The North is full of interesting people who live their dreams. Walt Humphries is certainly one; another is the intriguing businessman who picked up the claim when it lost its shine for the original owners.

Jack Walker didn't start out as a rock collector. He was raised in Westlock, Alberta on the Pembina River where, he says, "nothing around there even resembles a rock!" But from Alberta there's a road north, and in 1969 Jack took it. He settled in Hay River on the south side of Great Slave Lake and worked for Imperial Oil for four years. There, he discovered two things: the first was that he loved the North, and the second was that he could make a living working for himself.

Jack's years in Hay River and later in Yellowknife allowed him to build businesses from the ground up and to purchase others. He's obviously a clever and likable businessman. Over time he owned the Ptarmigan Inn in Hay River, a bus company in Yellowknife, a bottling company in Iqaluit, a hotel in Fort Simpson, and the Yellowknife Inn — *the* place to be in the capital.

He not only ran a business that fed, watered and lodged guests. The Yellowknife Inn was also, for some years, the home of the NWT Legislative Assembly. The Inn, located downtown on Franklin Avenue, was a hub in the city. Everyone knew it; everyone went there, locals and visitors. People visited in mukluks and spike heels, in wedding dresses and jeans. And more than all this, in the heart of the hotel was The Miner's Mess.

The Miner's Mess no longer exists, but back in the day, it was like a Tim Horton's (in the sense of being a relaxed gathering place for coffee), but more comfortable and with ambience. It was more personable and a one-of-a-kind adventure. There, you could have your own mug with your name on it. All kinds of things happened here. In 1984 when Yellowknife celebrated their 50th birthday, storyteller Jim Green recorded an album there. On regular days, though, it was where people came to get the news, the gossip, the hot tips. Miners, bush pilots, visitors, business people, writers, dignitaries, entertainers, church ministers and chiefs dropped by often, sometimes daily.

It was here that Jack first heard about the Acasta gneiss. "I was interested," he says. "There are old bits [of rock] in Australia and Greenland, but this was an outcrop. This was a piece of the Earth's crust that had risen to the surface, been pulled down again, surfaced and finally stayed. This piece hadn't gone down so deep that it was melted, so it has stayed as it was in the beginning. That's what intrigues me. There's no other value in it, no minerals that can be mined, for example. It's the age and the fact that's it's an outcrop that got my interest."

One day when Walt Humphries went for his cup of java at the Miner's Mess, he got into a conversation with Jack. The conversation turned to that claim on the Acasta River and within a short time, they had struck a deal. Soon after, Jack Walker chartered a plane north.

Twenty years later, Jack has divested most of his holdings in the North and retired to Alberta. He invited Bill and I to a property he's recently purchased with his wife Cheryl. We stood in the spring sun (this is in the Alberta foothills, so we also had snow for five minutes) and talked about his time in the North. We talked mainly about that one holding that he's not prepared to divest: the fifty-two acres on the north end of the island in the Acasta River.

Jack has done a lot of work related to his claim. It has now been upgraded to a lease on the land. He has tried selling a variety of souvenir-type items made from small pieces of the rock. One of the funniest to me was a clock. The irony of a time piece ticking away hours on the face — and in the face — of 4 billion years seems hilarious, although at the same time, sobering.

One of his business partners tried selling runes made from it. In fact, the mind boggles at the variety of items he has tried. Jack had the rock listed in the *Guinness Book of Records* in the mid nineties. He also sent a piece to The Vancouver Petrographics company that "thin slices" rock, dyes and photographs it. The image we saw of the gneiss was beautiful. Each individual mineral appears as a different colour and is enlarged; the picture reminded me of a stained glass cathedral window. In 2003, Jack gave away a piece that weighed around 6,000 pounds. He donated it to the Smithsonian Institute in Washington, DC.

Jack's manner is kind. He was generous in showing Bill and I his documents, photographs, samples and the mind-boggling assortment of items created from the gneiss. Perhaps it was looking through his boxes that reminded him of how much enjoyment this rock has given him. "I can see why Mark Brown is so excited about the rock," he said. "It *is* exciting."

Jack is not about to quit his dreams for the rock any time soon. He has "some ideas," he says candidly. His wife joins us as we say farewell. All the while we had been talking with Jack, Cheryl had been outside, walking along the river bank. She's creating a garden, she explained. She was out collecting rocks.

**Questions for Reflection**

1. What did you learn about the territory of the Dene in this chapter?
2. Geologists, trappers and hunters spend weeks and months on the land, sometimes in dangerous conditions. What did you learn about these people in this chapter?
3. What connection do you have to mining in Canada?

**The God Particle**

Around the world, scientists working in Geneva, Switzerland grabbed headlines in 2012 when they announced a discovery about the origin of the universe: *Higgs quest finds what looks like the 'God Particle.'*

International scientists had agreed that the Big Bang had occurred, but wanted to understand the *how* and *why*. The discovery of a new subatomic particle, the existence of which is fundamental to the creation of the universe, provided one more piece to the puzzle.

Chapter 5

# The Rock More Travelled

**Time is the best teacher;**
**unfortunately, it kills all of its students.**

Robin Williams, actor

Since its nomadic continental drifting around the planet, the Acasta gneiss has been settled near the top of the world for millions of years. Today, it is in the territory of the Tlicho Dene.

Since the earliest days of colonization, the government and individuals have poked and prodded, blasted and burrowed into this land, probing its secrets. The oldest part of the land, however, didn't come to geologic attention until the 1980s. Since then, it has become a celebrity. And, as you will see, pieces of the oldest rock continue to travel the globe, although not only on a slowly drifting continent. Nowadays it travels by bush plane, is crated in the cargo hold of a transport plane, nestled in a mail pouch, or carried in the luggage of passengers on jets.

When the Acasta gneiss was sent to Australia 1989 for age-dating, it experienced its first long haul flight. It was certainly not its last. And definitely not the most interesting.

Walt Humphries' principal interest in the rock is that it is a fascinating specimen. He didn't polish or shape it, but simply kept it as he found it. He was happy selling to collectors, but sending the rock to universities and museums around the world was even better since more people would get to see it. Generally, he would provide a piece of the rock if the institution paid expenses.

"I've lost track of all the places I sent it," he says. "Denmark, Switzerland, Britain, Japan and all around North America. Students of the Earth Sciences need good specimens to learn from," he says. Humphries also provided a piece of the rock for Jowi Taylor who wanted to unite Canada though stories and music.

**With A Song in his Heart**

It's hard to imagine what sounds surrounded the Acasta gneiss when it was underground. Maybe just stony silence. But for some billion years, when the gneiss was above ground, the music in the air was the natural kind. In the beginning the music was the prehistoric kind, comets crashing, wild wind, that sort of thing. Toxic rain fell and fell and fell, running over its face for centuries. Ice formed and melted, splashing and dripping down crevices.

Eventually new sounds rose around that gneiss. A sabre tooth tiger, perhaps, padding across the rock surface, crouching, breathing softly, waiting for prey. The trumpeting of a wooly mammoth searching for her baby could have echoed back from the rock wall. The sound of a dragon fly's wings likely whispered as it passed in flight. The rock may have heard early human voices as people camped nearby; their stories, weeping, music, laughter and crackling fires drifting over its face.

These days, we know that the sounds of trumpeter swans *en route* to breeding grounds and Arctic tern fresh from their annual migration from South America blend with the bawling of grizzly cubs looking for their mom. Human sounds infrequently invade the space: a float plane bringing geologists for brief visits, the splash of a paddle on the river; the crackle of camp fires, the pounding of a prospector's hammer driving a stake post into the meagre soil.

But now guitar music regularly drifts over a piece of the gneiss because of one man's wild imagining of Canada. A small piece is embedded in the fret board of Jowi Taylor's guitar, along with sixty-two other pieces of Canada.

When it appeared that Quebec might separate from the rest of the country, a lot of Canadians reacted. It was unthinkable that we could break up. Taylor conceived the idea to help keep us together through the magic of stories and music. He dreamed a guitar made of various materials that held stories for Canadians. He found George Rizsanyi, a luthier who had long practiced the art of building guitars with Canadian materials.The guitar has become a "passionate metaphor for Canada."

Taylor told me that when he was planning the project, he was looking for materials to go into the guitar that had a good story behind them. For example, there is slice from the Golden Spruce on

Haida Gwaii, a piece of an original seat from Toronto's Massey Hall and some Sudbury nickel. There is mammoth ivory and a piece of gold from Rocket Richard's Stanley Cup ring.

He was more than pleased, he said, to receive a piece of Acasta gneiss through Walt Humphries.

"Certainly hearing its story and its role in the territorial mace in the NWT Legislature made me think it would be an appropriate item. Each and every piece in the guitar gave me pause when I first touched them; I considered where they had come from and their destiny as part of this project. I actually got to wear Maurice Richard's Stanley Cup ring before we cut the little piece of gold from it; that made me think about his place in various histories... of course, with the Acasta gneiss it was really this moment to think of what an extra-ordinary thing it was to be holding something so tremendously ancient and how it would now have this different sort of life with music resonating through it."

The music that passes over this particular rock piece is made by amateurs and professionals alike. Taylor's book, *Six String Nation*, tells the unique story of his uncommon dream. The guitar made its debut on Canada Day, 2006 at the national celebration in Ottawa. Taylor tours Canada regularly with his aptly named guitar, *Voyageur*.

Guitar music is one form of music wafting over the gneiss; the gentle rustle of pebbles and multi-lingual governmental debates are another.

**Down the Road, if there was one**

When Nunavut separated from the NWT and became its own territory in 1999, the NWT needed new symbols to represent its new reality. Among them was a new territorial mace which opens each session of the Legislature. The mace is kept in Yellowknife, 300 km down the road, so to speak, from the Denàdzìideè/ Acasta River.

Artists Bill Nasogaluak, Dophus Cadieux and Allyson M. Simmie created the new mace representing Inuit, Métis and Dene cultures. It features weather, the land and ten of the languages spoken in the territory. They incorporated beadwork, metalwork and porcupine quillwork, and even stones. A pebble from each of the thirty-three NWT communities was placed inside the mace so that when it moves, the shifting pebbles create a sound similar to that of a rainstick. This sound, the literature says, "is to represent the united voices of the people."

This unique mace rests on a stand of white marble, with carvings to represent the Mackenzie/Dehcho River, Great Bear and Great Slave lakes. It is adorned with gold and silver Mountain Avens, the territorial flower. Beneath, are clusters of Acasta gneiss.

Shortly after the gneiss made its appearance on the mace, the most visited museum in the world came calling.

**Off to Washington**

In 2003, the Smithsonian Institute in Washington sent a plane to the site. James Pepper-Henry, a director at the museum explained that the museum hoped to extract the rock without disturbing the site too much. A Dene delegation was on the site for a ceremonial blessing as the museum staff readied the stone for travel to Washington, DC. When The Museum of the American Indian opened the following year, Tlicho and other First Nations peoples were present to offer prayers and ceremony.

The gneiss is one of four rocks in a stone display representing Aboriginal peoples from the four cardinal directions. The museum was designed by Douglas Cardinal, an architect whose Métis and Blackfoot ancestry inform his buildings. Among other distinctive public buildings, he has designed the Canadian Museum of Civilization in Gatineau, Quebec and the First Nations University of Canada in Regina, Saskatchewan.

The gneiss seems to act as a magnet for people who love rocks and appreciate Earth history. Steven Schimmrich is a geologist in New York who writes a blog called Hudson Valley Geologist. He travelled to Washington in 2011 when he learned that he could actually touch a piece of the gneiss. That kind of interest is part of the reason the gneiss became part of a rebuilding project on the other side of the world.

**Peace Offering**

In 2006, a five pound piece of gneiss became part of a peace offering in war-torn Afghanistan. According to Northern News Services, "a little piece of the NWT is helping Afghan geologists rebuild their people's interest in Earth Sciences." The British Geological Survey had partnered with the Afghan Geological Survey and together were helping to modernize the Geological Museum in Kabul, a peaceful project. Donna Schreiner, outreach geologist with the NWT Northern Geoscience Centre in Yellowknife says that every year there are requests for northern rock samples.

**Here, There and Everywhere**

Many rock collectors have added a piece of the gneiss to their specimen stockpiles. Apparently, none of the geologists or claim holders have kept track of where they've sent it. I was intrigued to stumble upon a reference to the gneiss in Ross Laird's book and enjoyed his poetic reaction to it. The scene he describes takes place in British Columbia. It is in his book, *A Stone's Throw: The Enduring Nature of Myth.*

Laird writes, "[He] received the gneiss as a gift when he worked as a geologist for the Canadian government, and he counts it among his most prized possessions. I can see why: ... at 4 billion years of age, such an artifact is astonishing — older by far, than the dinosaurs, older even than the earliest fossils. It invites questions and ruminations. It encourages within me a sense of my own insignificance. But also, like an anchor offered to the drifting mind, it provides a concrete means of joining with the past. Four billion years is a difficult span to imagine. Measures and metaphors can capture the breadth of that time, but not its depth. Here, though, cradled in my palm, is four billion years

"Looking closely, I glimpse the detail in each of the black bands: whorls, curves, marks like tiny swirls of ink. Dark tendrils extend above and below each striation. Here and there the white breaks through, truncating a long skein of black. ... it begins to resemble a text ... black on white, linear, structured with letters and punctuation. The similarity is unmistakable. 'In the beginning,' says the Hebrew legend, 'the Torah was written with black fire on white fire, lying in the lap of God.'" It's hard to say if the geologists in Japan waxed poetic like this, but I love their idea of a wall display.

**Hitting the Wall in Japan**

One interesting request from Japan arrived in Walt Humphries' mail box one day, from Tokyo University. "This was in the days when Japan had a lot of money [before the Tsunami and nuclear meltdown in 2011]. They were building a geological wall to show the Earth from the beginning of time. They wanted to have as their first piece, the Acasta gneiss. From there, they added ancient fossils, volcanic rock and so on. It would be quite a wall to see, several hundred feet long. They went all over the world in search of ancient and interesting specimens. They came here themselves and hauled away about four tons of it," he says.

**Hawaii**

John Ketchum, Senior Geologist at NWT GeoScience office in Yellowknife said that in 2009, while on vacation in Hawaii he took a piece of the gneiss to the Hawaiian Volcano Observatory. "I was amused by the concept of giving a piece of Earth's oldest rock to a group of people who study and regularly watch the formation of Earth's youngest rocks." Personally, I found that thought romantic. But perhaps the prize for romance should go to Mark Brown, the owner of the southern part of the Acasta gneiss claim. For that story, recall the biggest wedding in the world in 2011.

**To Royal London**

When it was announced that Prince William and Princess Kate would travel to the NWT on their honeymoon, Mark Brown went to work. He sent a sizable piece of the gneiss to be cut and polished into a fist-sized sphere. Then he began jumping through the official hoops he encountered, seeking to make his offering an official wedding gift on behalf of the Northwest Territories.

A meeting with his member of the Legislative Assembly resulted in a list of further hoops and security clearances. In the end, the hoops became too many and the time too short. Undeterred, Mark went to Plan B, and gained permission to set up a display of the rock, the sphere and an old British flag he'd saved from the days he worked in an antique and curio shop. He set up a table at the Visitors' Centre, close by the royal walking route from Somba K'e Park at City Hall to The Prince of Wales Northern Heritage Centre.

It was a clever site to choose, and Mark received many residents and visitors who had come to welcome the regal guests. Alas, the prince and princess were royally rushed along their walking route and could not stop to receive their gift.

Mark is many things, but he is not a quitter. After the royal hoopla ended, he packaged his precious gift and sent it to the Governor General. There it was cleared and sent to London as part of Canada's weddings gifts.

Mark took the time to ferret out a complete list of gifts from Canada given the royal newlyweds. Through a Toronto Star weblink, he achieved this. The gneiss appears on the official list as "rock sample." Mark learned that the NWT premier had given the couple diamond jewelry. He mused that while NWT diamonds are unique in the world, it's likely this couple doesn't really need more jewelry.

And then he spotted what our prime minister had given them: a blanket, a photograph, magazines, jackets, a bottle of whiskey and a Canadian flag. "A blanket and flag? A bottle of whiskey? When he could have given them a gift that is unique in the world!?"

Mark laughs and shakes his head at the wonder of it. Then he carefully unfolds his thank you note from Buckingham Palace and rereads it.

Mark cannot reread the original Treasure Map, as he calls the map marked with the gneiss location, because a cleaning lady at his hotel inadvertently threw it into the garbage. But that map is embedded in his soul now, as if it were etched in stone.

**Questions for Reflection**

1. What surprised you about the stories in this chapter?

2. In your part of the world, how old are the rocks? How are they primarily used?

Barren-ground Caribou

Chapter 6

# Treasure Map: Rocky Road to the Gneiss

**Today you are You, that is truer than true. There is no one alive who is Youer than You.**

Dr. Seuss

Like the others connected to the Acasta gneiss, my nephew Mark Brown has made his living in the North. He's worked at several of the mines above and below the Arctic Circle. Whether the temperature is forty below or twenty-five above, Mark loves this northern land and loves the physically demanding work. And that's how it is that he ended up with a passion for this rock. It's a story straight out of Monty Python, a book of fairy tales or ancient myths, or maybe a combination of all three.

As Jack Walker, who owns the claim on the north end of the site has said, Mark is excited about the rock. Friends and relatives will attest to the fact that he can talk about little else. His claim on the south end of the island in the Acasta River is a little smaller than Jack's, due to the fact that some of it is in the river itself. Mark staked his claim in 2008 on crown land, and restaked it in 2012 to correct one post.

Geologist Gina Marie Ceylan said of the Acasta gneiss that, "heat, pressure and fluids have moved through this rock and changed it, but it's still essentially the same rock. This rock has been through so much. And it's analogous to human experience. In life we go through all this stuff." Mark knows about going through stuff, too. His journey to and with the rock has been a road less travelled. And rocky, you could say.

The way Mark tells it, he stumbled upon his treasure map. That accidental find set him on a quest to connect with the story held inside that rock, to try to understand it and to let the world know about it, too. Mark is a dreamer of dreams; like most dreams, they seldom fit into the proverbial box.

It's been said that the North attracts interesting characters; Mark qualifies. Mark's work in the North came about as a result of having a drink at the famous old bar in Yellowknife, The Gold Range.

"It was Christmas in 1984 and I was in town visiting family. A guy at the next table asked me if I wanted to go diamond drilling. I said, "Sure." Next thing I knew, I was underground on the tundra looking for gold, hung over like a palm tree.

"I stayed four months; never saw the sun for three. I always had a job after that." Mark says he loved it underground. I suppose that since it's pitch black above ground in winter, it doesn't make too much difference if you're above or below the surface. "It was like beating winter and conquering fear at the same time," he says. "People in the camps form a tight community out of necessity; everyone needs each other. It's life and death. You really do have to be your brother's keeper."

Mark says that the Internet has changed camp dynamics to some extent; there is less conversation. "But everyone still competes in storytelling, especially animal and weather stories. We look for the biggest, worst, hottest, coldest, smallest or most terrifying in our stories." There are a lot of bear and wolverine stories, too. "Just about everyone has a wolverine story. Sometimes it's freezing and you open the outhouse door and find a wolverine in there. Other times, it's a hot sunny night and you open the tent and there's a wolverine standing on your sleeping bag." Wolverines, according to the dictionary, are heavily built, short-legged carnivorous mammals, native to the tundra and forests of arctic and subarctic regions. What it doesn't say is that they can easily kill you and are smart enough never to be caught in traps.

"Mining exploration is unpredictable at best," Mark says. "We're working adventurists and we thrive on it. Sometimes going into the bush is like serving a sentence and I plan my time there carefully until I can get out." His love of reading has been helpful for his own education, which formally ended after grade eleven. His tastes, like him, are eclectic: history, world religions, biography, geology, psychology, biology, geography, practically any nonfiction.

"But mostly," he says, "I look at going into the bush as if I'm going to a health spa. There's fresh air, exercise, good food. And there's no alcohol." Alcohol has been a problem for Mark.

For single people working seasonally or in the bush, housing is often an issue. For Mark it's been complicated. He's in the bush about six months a year. There, he's housed, but in town it's been a variety of roofs over his head. He's lived with various family members on and off, rented apartments and houses; stayed in a travel trailer, a house boat, a bus, a tent, the Salvation Army hostel, a children's play house. For the past few years he's lived at a hotel on Franklin Avenue in the northern capital city. "Maybe if I ever make some money I'll buy a farm on *the* Rock [Newfoundland]!" he laughs.

Neither Mark nor I are born northerners; I moved to Yellowknife in 1970 and he in 1984. We were both raised "outside," as northerners like to call people from south of the sixtieth parallel.

"All in all," he reflects, "It's been a mercurial ride. I wouldn't want people to think it's all been a lark; at times I don't have a pot to piss in. Sometimes," he adds, "I am *this* close to hitting rock bottom."

Certainly his sense of humour, creativity and ultimate love of life have seen him through the years and wild adventures that would have defeated others.

Mark was born on the longest day of the year in 1965 and raised on a small mixed farm near Orangeville, Ontario. His mother, Marty, was a homemaker and community volunteer; his father worked away from home during the week selling automobile parts. Mark's family was then, and continue to be, deeply involved in amateur theatre and sports. Growing up, Mark was a valued lacrosse and hockey player. When Mark's father Hugh died suddenly of a heart attack at a Toronto Christmas party in 1981, the family's world caved in. There was no warning, really. Mark was sixteen years old. He does not like to speak about that time.

Within two years, his mother and younger brother, Matt aged twelve, had packed up the household and moved to Yellowknife. They stayed in our home while Marty looked for employment and a home. Mark's sister Michele was at Brock University in St. Catherine's, Ontario. "I didn't want to move North," he recalls. "I didn't want to go to school. I didn't want to move in with relatives, although I could have." Death is rarely fair and often cruel.

At first, Mark remained in Ontario and lived with family friends and later, moved in with Laurie, whom he describes as "a prostitute who was the smartest entrepreneur I've ever met. By the time she was 18, she owned three houses and had a body guard."

Mark worked in retail, sold fire wood, then ran his own chimney sweep business for a time. Eventually, he headed out to see what might happen north of sixty. In the mid 1980s, the price of gold was high and there were lots of jobs. "I could pick and choose where, and if, I wanted to work," he said.

Mark worked above and below ground for five years at various camps all over the NWT, above and below the Arctic Circle. He took occasional breaks for travel in Europe and southern Canada, for other work and for visiting relatives. On one of his forays to Alberta, he got a balloon pilot's license.

Mark hoped to give rides to people of course, but his bigger idea was to paint *Give to the Heart and Stroke Foundation* on the balloon and donate a portion of his passengers' fees to the fund, too. (Unfortunately, the balloon he purchased was a lemon, but that's *another* story.) He sold his truck and balloon and headed north once again. Soon after his return, he got lucky — or so he thought. He received a sizable inheritance from a relative.

He took his windfall and moved to Costa Rica in 1990. He stayed almost three years. There, he started a publishing company that produced car trader magazines in Spanish, a language he doesn't speak. He had a grand time, he says, but "came back addicted to just about everything." When he returned to Canada, he entered rehab in Vancouver for the first, but not the last, time.

Around this time, Mark's paternal grandfather developed prostate cancer. Mark returned to Ontario where he worked for a furniture mover in Newmarket. He valued the time he could spend with his three grandparents; they appreciated his visits, too. There were few other relatives left in Ontario by that time.

As a furniture mover he learned that people sometimes leave belongings behind. "It occurred to me that a nonprofit organization could gather up that stuff and auction it off to raise money." In 1996 Mark attended the auctioneer school of Woodstock, Ontario and for a time, worked with the Diabetes Association to try to put his plan into action. "I offered to run the auctions. They thought it was too good to be true." Unfortunately, the relationship didn't develop further.

Mark grew restless and headed back West in 1998. He stayed with my sister's family in rural Saskatchewan for a while, and with our family, now living in Calgary. He also worked in the Alberta oil patch for a short time. I remember that Mark was trying to figure out how to develop a more healthy life style during this time. Eventually he hit on what seemed to him a fool-proof plan. He would move to a kibbutz in Israel.

In preparation, he returned to detox in Vancouver and then flew directly to Tel Aviv in February. "I'd always wanted to see the Holy Land," he recalled. He had permission to stay one year. He stayed a year, plus the month he was in jail.

Mark found his way to a kibbutz on the Jordan River. "The guy in charge said, 'You're a Canadian, you know about adventure.' He sent me on a white water rafting course with the idea that I would become a guide for tourists. I was a month into the training when someone asked if I was allergic to bee stings. I am. That was the end of it. They terminated me."

Mark says he enjoyed Israel and took in many of the sights, the markets, and drank tea with Bedouin people. His employment varied. He tended bar on a Red Sea casino boat, drove a taxi, worked as a mechanic's assistant and in a scuba shop. He was head barbecue chef in a beach restaurant, and was part of the entertainment on a nude beach. That last was "one of the few jobs I've had where I looked forward to work every day. The party ended at 2:00; I was up at 6:00 am slicing cucumber and tomatoes."

But then, he says, "I got a game misconduct." He was charged with "fighting, malicious property damage and inciting a riot." He ended up in the Ramallah Penitentiary.

"The inmates were friendly," he recalls. "We could only have three cigarettes per day, but the Palestinians threw cigarettes over the fence for me." The month was long since it was spent sleeping in tents shared by twenty people, and lots of bed bugs. "It wasn't Three Star," he laughs.

True to form, he made the best of it. Since gambling was forbidden, playing cards weren't allowed. Mark and his friends made some from cut up cardboard boxes. "We played Crazy Eights for cigarettes, even the Romanians who couldn't speak English. One day a big guard said, 'No gambling here!' We said, 'Oh fuck off, You wouldn't give us any cards, so we made up our own game.' He didn't pursue it.

"By the end of the month I had boils on my skin, I hadn't shaved and I'd lost teeth in a fight. I looked pretty rough. They came to get me at 4:00 a.m. and took me right on to the plane in cuffs. The flight attendant says, 'You're Mark from Canada?' It was like an announcement." He laughs at the memory, incredulous that this is his life. "My seat was in the tail of the plane, so I had to walk the gauntlet."

He lived in the Vancouver area after that, sometimes with his brother, and was in and out of rehab. In 1999 his maternal grandfather died. Thinking about the little time he'd have left with his father's father and mother's mother, he returned to Ontario.

Mark moved back to Newmarket, across the street from my mother. I'm grateful for this; mom was grieving not only her husband, but she had moved to a retirement facility. She was grieving her home, garden and land. She was grieving her failing eye sight, too.

"I'd get a pretty good buzz on," Mark says, "and go over to see Grandmother. We'd put on her favourite music, Andrea Boccelli, and dance. It was the privilege of my life. She loved to dance. I've had some pretty good times with my grandparents."

It was around this time that he came up with another way to raise awareness and funds for the Heart and Stroke Foundation. Mark had the idea that he would set a bath tub outside and fill it with beans. "Beans," he explained, "are good for your heart." Mark planned to set the *Guinness Book of Records* for sitting the longest in a tub of beans. He had read that the record was forty-two days. "I figured I could do it for forty-three." In retrospect, he wonders if he'd have really done it.

At the time, though, he figured that people would remember the rhyme. (In case you've forgotten: Beans, beans, musical fruit, the more you eat the more you toot, the more you toot, the better you feel, so let's have beans for every meal.) He thought that people would laugh and then make a good donation to the Heart and Stroke

Fund. He went so far as to approach pork and bean producers, but that vision became a blow-out, so to speak.

Mark attended the burial of his grandmother in 2003, then moved to Port Credit and sold antiques and curios. "I lived there with my grandfather for a year," Mark recalls tenderly. "In all my life, I've never met a gentleman like him.

"One day Grandfather phoned me at work and said, 'I think I need your help.' I took him to the hospital and saw him settled into his room. As I sat with him his face suddenly became illuminated. I believe that he saw my [deceased] grandmother, waiting for him."

Soon after his grandfather died, Mark decided that the North was the best place for him. He returned to Yellowknife, and back to the camps where he was diamond drilling and expediting. His life took on a familiar rhythm: too much alcohol in the city, sober and healthy in the camps. Mark had been back working in the North two years when the treasure map, as he calls it, fell into his hands. Three years later in 2012, I interviewed him at the Discovery Inn where he lives, in downtown Yellowknife.

Before we begin, he shows me a beautiful eagle head and various pieces of jewelry set in silver that he has had carved from the gneiss — rings, pendants, cuff links, earrings. He also has highly polished orbs; they feel good in the palm. We examine rocks with various striations and colours, some polished and some rough. Rocks dominate his room. Shelves, table tops, end tables — there is little extra space. We settle in and then Mark begins what he does well. He told stories.

"In the beginning," he says, "there was this rock. Every society needs a platform, something to stand on. I got a rock to stand on when nothing else really made sense. There are all these religions, but I don't know what's better than the oldest known rock. It's a place to put your feet down. Solid."

Mark reaches into his pockets and palms a meteorite in his left, the Acasta gneiss in his right. "Balance," he says. And then he sips his vodka, leans forward on the table and begins to tell the story about the map.

"I was working for a South African Company on Victoria Island, north of the Arctic Circle. They were looking for diamonds. Our crew was hired by an expediting company to build exploration camps — building a little village, really — cook houses, outhouses, bunk houses, everything. I was minding my own business, been there about four weeks. It was February, 2008. It's forty below, top of the fucking world, freezing our asses off. We had to dig deep down through sixteen feet of snow to get to land, and build the tent frames down in a hole.

"They're walled tents roughly 14 x 18 feet, made of canvas and plywood. So the idea was to get ready for the exploration crew coming in the spring. There were about eleven of us. We all ended up sleeping in the kitchen because it was the only completed shelter available. We slept wherever we could put our sleeping bag. I love this business, it's so unpredictable!" Mark laughs at the memory.

"Before you go out to the camps, you get all these fine plans ready. You're sitting in Yellowknife with all these lists. But after you leave town, you might as well throw those plans out the window! Nothing really works out like it did on paper.

"Unpredictable or not, I love it. I love the life. Where else can you wake up in the morning and see a herd of muskox?" Or, of course, wolves, those amazing wolverine, migrating caribou, literally thousands of birds on the wing and pristine sky, water and land?

"Anyway, we got the tents built," Mark continues. "And we're loading up the kitchen and the geological stuff, getting everything ready for the team's arrival. We're sorting boxes, maps, charts and it so happens that a map falls away from the rest of the stuff. I was working with Ron, an Inuit fellow from Cambridge. We look at it and see that all the active and inactive camps in the Territories are marked. We unfold it, look it over and start naming where we've worked and with whom.

"I notice 'Acasta River gneiss — the oldest known rock on the planet' is marked. I couldn't even pronounce gneiss, but I was curious about it. I tried to play it cool, but I was intrigued.

"I went to see the geologist; a South African. 'What's the oldest known rock on the planet?' I asked. He said, 'The Acasta gneiss. You guys won it for that.'

"My mind lit up. I thought it was interesting — and a mind blower! I wondered, Who owns the claim on that? And, if you had it, what would you do with it?"

Mark was raised in a family that was neither rich nor poor, but like most everyone does, he's played the *What if?* game, as in, What if I won a million dollars? What if I had everything? What would that mean? He thought about Mick Jagger of the Rolling Stones. What if Mick heard of the rock?

"Mick could have anything. Diamonds. Cars. Gold. But, does he have the oldest rock on the planet? Would he want a piece? My mind went everywhere!

"After I returned to Yellowknife I went to the Geoscience Office of the federal government. Dr. Stephen Geoff, the senior geologist there was generous and kind. He made sure I knew the answers to every one of my questions."

The more he learned, the more passionate Mark became. He believed that this rock was something big, something important — not only to him, but for everyone. Within months, he had spent time with Senior Geologist John Ketchum and prospector Walt Humphries, the original owner of the claim. He learned as much as he could, including that NASA and the federal government retain rights to harvest rock from the site. In his mind, that only added to the value and his interest in the site.

Throughout August and September, he couldn't get that map out of his mind. He kept asking questions, talking to people about it. One person he talked to was a helicopter pilot, who said, "You know, you're on to something good. You could sell small pieces of it inside a pyramid." He discussed going into partnership with Mark. But before that could happen, he had a heart attack and died, aged 40. Shortly after, Dr. Geoff also died of a heart attack. Mark was shaken. His own father had died of a sudden heart attack when Mark was only sixteen years old. But the dream of staking a claim on that rock did not die.

He dreamed of creating a company to sell jewelry, orbs and raw pieces of it; he couldn't help but name it Rock of Ages, NWT. He says he doesn't remember singing the 1763 hymn of that name in his childhood church, but he certainly remembers singing along to The Band's rock album in the 1970s. According to Mark, that was music to grow up by; Bob Dylan, Robbie Robertson and all the rest of that era.

On October 16, 2008 Mark chartered a Cessna float plane with a local firm, Air Tindi. The trip took ninety minutes one way and cost

$2,000. He says he didn't really know what to expect. "I had a rinky dink little camera," he says, "but it was impossible for me to shoot and do the staking at the same time.

"The pilot who flew me up suggested I speak to Terry Woolf, a prominent filmmaker in the North. Soon after, I read a feature in the newspaper about Western Arctic Moving Pictures (WAMP), a local co-operative that Terry is part of. What stuck in my mind was that the article said that if someone had a good idea, even if they didn't have much money, that WAMP would try to find someone to run with it." Mark made the call.

Soon Mark had an appointment with Jeremy Emerson, the executive director of Western Arctic Moving Pictures. "Jeremy loved the idea right away," he says.

It was January, 2010. Hours of daylight are scarce at that time of year. It was 48 below zero when Mark's next charter landed on the Acasta River. This time he had a film crew and the plane had skis.

**Questions for Reflection**

1. What value do you and your friends place on storytelling?
2. In a usual week, where do most of the stories you hear originate? (For example, television, Internet, newspapers, theatre, visits with friends.) Has this changed over time?
3. Mark is a character who teaches the author to see the world in different ways. Who in your life does this for you?

After the sun returns to the Arctic in spring, Mark Brown and friends might do a little top of the world snow golfing after work.

We are the generation that has come to witness the beginning of the universe. The common origin story tells us that we are primarily cosmological beings.

We are the first generation to have the experience of seeing the universe and seeing ourselves as an expression of it.

Brian Swimme in Calgary, June 2012

Chapter 7

# Bags of Frozen Rocks

**I've had university profs look at my price on a piece of the rock—$150. They asked, 'Could I get one for $120?' I couldn't believe it. I thought, *You have a PhD and I might live in a cardboard box and you want to save $30? Well, then, if it's that important to you, I'll make sure you get one.***

Mark Brown January, 2012

Our daughter Andrea and her son Tristan walk with Bill and I up the Bush Pilot's hill in Yellowknife's Old Town to meet Jeremy Emerson. They know each other of course; Yellowknife may be a capital city but the Old Town is small. The shack Jeremy shares with Roberta, his school teacher partner, is just above the famous Wild Cat Café. The climb, which includes vintage wooden stairs clinging to the Precambrian rocks, is worth it. He's got a million dollar view of Back Bay. On May 21st, the ice is just beginning to break up.

Andrea and Tristan continue their walk while Jeremy gives us a tour of his yard. This includes a greenhouse, garden, a dilapidated children's playhouse (in which Mark lived for a couple of months), and a quick tour of their cozy home which features an indoor honey bucket, a home office and funky art. It is one of the original Yellowknife shacks and it has a good vibe. But the sun is warm, the birds are singing and we decide on an outdoor interview near the fire pit.

Jeremy grew up in Fort Smith, NWT. Twenty-five years old, a recent graduate of Lethbridge University in Alberta, Jeremy returned North and settled in the capital. It wasn't long before he caught the attention of Mark and began receiving emails from him. "He was passionate about a rock," recalls Jeremy, "and he wanted me to make a film about it."

But it wasn't the rock that prompted Jeremy to agree. It was Mark. He recalls the first time they met.

"He emailed that he wanted someone to go out to this island and make a film. I thought I could send his request around to our membership [of northern filmmakers] and see who was interested, but Mark thought I was the guy to make the film. He arranged to come by. He's brought these big cue cards [bristol board] and a map, and he starts setting up in the office. He's rolling out the map and he doesn't say much at first, but then he starts pitching it, almost like a Dragon's Den pitch."

The Dragon's Den is a popular Canadian television program featuring five millionaires — the "dragons." One after another, people appear before them, trying to interest one in bank-rolling their business project. It is, in turn, fun, interesting and heart-breaking to watch. As it turned out, Mark's pitch was a rehearsal to actually meet the Dragons.

Jeremy recalls, "I'm thinking this guy is really strange; is he for real? He starts, 'In the beginning... the oldest known rock in the world ... this is the foundation...' He was talking about sending rocks to Bob Dylan and the Rolling Stones. He wanted me to film his trip to the island. He asked, 'How much do you need?' and pulls out some cash. I wasn't even sure I wanted to do it at that point."

Jeremy had never heard about this gneiss or concerned himself with Earth sciences before. But Mark didn't let it go. He took Jeremy to meet John Ketchum at GeoSciences, sent him information, links, websites, names of people, connections and all kinds of geological articles.

It worked. Jeremy became curious. He began talking to people about the rock, a film, about geology and the age of Earth. Interestingly, Jeremy was raised in a religious family that doesn't believe in evolution. "The four billion years doesn't really work for them," Jeremy says thoughtfully. But he works it out a different way. "Evolution obviously occurs," he says, "but that doesn't mean there's no Creator."

Jeremy didn't know what he was getting into, but he liked Mark. "I had no idea that it would last this long, or that I'd develop a friendship with Mark. He was living at the Sally Ann [Salvation Army Hostel] when we met. He lived here for a while, too, and that was really cool. He's a good story teller; he has a natural skill for it. That helped me stay curious and interested in his story. I'm more interested in Mark than I am in the rock."

Bill and I take our leave and climb up to the Bush Pilots' monument above Jeremy's home. We rest a moment enjoying the view of houseboats waiting impatiently for Great Slave Lake to become ice-free.

**The Film**

It's not a simple thing to shoot a film in the North, at least not in winter. Frozen cameras, ice fog, long hours of darkness, breaking ice, grounded aircraft, little things like that make it expensive and time-consuming. Making sure that the pilot, sound and camera people were available on the same day was another hurdle.

"Because of one thing or another, we had five cancelled trips," Mark says, "You can't fly in there during break up or freeze up of course, and things slow down over Christmas. Come 2010 I was chomping at the bit to get going.

"Finally we got there, just after Christmas. It was 48 [degrees] below, cold as hell, we just rolled with it and got the job done. We had to pick up rock samples. The snow was up to our waist and we had one pair of snow shoes. And of course, we had to keep the camera batteries warm, so everybody had them strapped to their chests. All in all, it turned out really well."

**The Vancouver Olympics**

"When Jeremy got back to Yellowknife," Mark says, "he edited the film and then posted it on YouTube. Next the government of the Northwest Territories picked up on it. They were looking for stuff for the Northern House at the Vancouver Olympics. They loved it right away and wanted it. I said to Jeremy, you look after the film. If they pay you, you take the money, I don't care. But the good old government they didn't even invite me to the party, they just wanted the film. I had to scramble about and talk to a government person about that. They said, 'No, you aren't invited.' I told them I wanted to sell rocks there, but the guy said they didn't have shelf space and all sorts of other bureaucratic stuff. I said, 'I'm going anyway,' and he said, 'Well, you can do what you want.'

"I went down with my bag of rocks and stayed about 10 days at a trippy old pub downtown. Basically what I got to do at Northern House was watch the film on the big screen. I couldn't solicit to sell rock, so I had to wait for people to see me, see the film and put two and two together. If they did, we had a conversation. It was really shitty. It could have been good exposure for me. One of the funny things though is that

Digawolf [a northern musician] was upstairs entertaining people, and it's him playing on the film, too. That part was fun." Mark sold two rocks.

A reporter for the Globe and Mail newspaper saw the film and later flew to Yellowknife to write a story about the diamond mine. He phoned Mark on his cell and asked for an interview. Mark was still in Vancouver. "I borrowed money for a plane ticket back North and talked to him for five minutes. The photographer fired off twenty photos and then they were gone."

"Two weeks later I was at Jeremy's, living in the playhouse. It was a hoot — about 4 by 8 feet, no insulation, no power, just an extension cord and a mattress on the floor.

"I'd been partying until about 4:00 in the morning and the reporter phoned around 8:00 wanting to do a full length interview. It was quite hilarious.

"He thought it was really quite an incredible story that anyone could actually live in this little shack. He asked how I charged my cell phone. I told him that I go to parking lots. They all have power to plug in your vehicle in the winter to keep them from freezing."

Mark also told him that he'd eat squirrels if he had to, in order to make his rock business work. It made good copy. The piece ran in April, 2010, front page. Mark sold dozens of rocks and jewelry pieces through his website shortly after.

"The Yellowknifer newspaper, CBC and Above and Beyond magazine wanted interviews, too. It was all kind of happening for me at that point. And then in September things got very quiet and I realized that I'd have to do the direct marketing approach. Around that time I learned that the Dragon's Den was coming to town to conduct interviews for their show. I'd been watching that for years. I was terrified just imagining being on the show. It would be good exposure but pretty scary."

In January 2011 Jeremy and Mark chartered a plane for another trip up to the Acasta River, but they ran into ice fog and couldn't land. That run cost $3,500. But maybe that's what gave Mark the courage to shoot for the stars. He applied to audition for the Dragons.

**The Dragon's Den, CBC Television**

The Dragon's Den audition was March 2011 at the Yellowknife Inn. "It was pretty nerve wracking," he says. "I wondered, How should I present myself? I started with four or five shots of vodka. I had my rocks wrapped up in towels from the hotel where I was living, and one piece in an old gray sock. I carried them over my shoulder in a canvas bag." He also took along Jeremy and a sound technician.

"For the actual audition there was Molly, a producer, a reporter for the Yellowknifer News and me. It was kind of primitive. I had my rolled up map and the rocks. I walked in and said, "Are you ready for the story? This is the product and this is what I'm doing. And then we started chatting away. When I left the audition I thought that they were going to call me. They did.

"They emailed saying I'd been selected and that I had to pay my own way but that I'd receive $1,000 toward expenses. I was to stay at the Royal York Hotel in downtown Toronto. I didn't have any money, but a friend loaned me enough for an airline ticket and my cousin made the Royal York reservation for me. This was a couple of weeks after the audition."

Mark took forty pounds of rock, a map, a regular axe and his pick axe and caught a ride to the airport. He had a two hour flight to Edmonton, a two hour wait and then a four hour flight to Toronto. But in Edmonton, the plane "went mechanical." He didn't arrive in Toronto until 6:30 am. "I was dog tired," he laughs. He was due at the CBC station at 9:00 am.

"I got my stuff, jumped in a taxi to the Royal York; but they wouldn't honour my hotel reservation because I didn't have the actual physical credit card. I had $190 in my bank account; not enough for a room. I explained everything to the guy at the desk, told him about the Dragon's Den and that I only needed two hours sleep and a shower. He sent me to another hotel around the corner. So, I'm on foot with 40 pounds of rocks in a canvas bag, an axe and pick axe in downtown Toronto. I paid the last of my money for the room; now I'm completely broke. I slept for an hour and a half and showered. I was just getting to CBC when my aunt phoned to wish me well. The last thing I expected was a phone call. My nerves were shot. I nearly went through the roof when the phone rang."

Mark met Molly in the Dragon's Den studio, had a tour then went to the lounge to wait along with Mark Donnelly, the opera singer who performs the national anthem for the Vancouver Canucks hockey team." He waited until 4:00 o'clock that afternoon.

I can't imagine what that wait would have been like. It reminded me of the first time I was on a television talk show about a book I'd written. I'd bought the outfit, fixed the hair, had my lines down, fought the traffic, found the studio. And then they wanted me to *wait?!* But for Mark, he was waiting to see five millionaires who were going to toast him, one way or the other.

"Eventually another producer asked how I was doing," Mark says. "She says, 'some people have difficulty with Kevin O'Leary. What happens if he gives you a hard time?' I said, "I came here to sell rock. If he gives me a hard time I might go over and punch him. A big *kafuffle* might happen. I might get arrested. The media might have a field day, and guess what? I might sell a lot of rock!" She never said a word, just turned and walked away. The funny thing is, a lot of people who watched the show said, 'Wow. Those Dragons were really nice to you!' I watched the show and agreed. A lot of other people do have a hard time. But when Kevin started to say something .... the other Dragons wouldn't let him get a word in. That was funny."

Mark's segment aired on the Internet version of the show in April 2012. The thirty-five minute conversation was edited down to five. What you won't see if you watch the video is the shot of vodka Mark swallowed when he really should have had lunch. You won't see Robert Herjavec pulling cash out of his pocket to buy a rock, or hear the conversation about why Mark loves his work in the North, the questions about fishing or the sideways nods from Jim Treliving and Arlene Dickinson which boosted Mark's confidence. You won't see his axe and map falling over with a crash. And when the cameras stopped rolling, you also won't see that the cameramen and technicians pulled out their wallets to purchase rocks.

When it was all over, Mark said, "Jim was like a grandfather or coach. I felt that he was on my side from the start. He's been to Yellowknife a few times, there's a Boston Pizza here. He helped me."

Of course, he wasn't calm standing in front of the Dragons. Who would be? When Arlene asked if he was the only person who had a claim on the rock, Mark answered Yes. But of course, Jack Walker has a claim, too. I surmised that Mark thought she meant, "Do you have a partner?" I asked him about that later, he said simply, "I misspoke. I was nervous."

"When Dragon Bruce Croxon made the offer, Arlene suggested that I run, not walk, to accept it." It remains to be seen what will come of the relationship with Bruce Croxon. He offered to meet Mark's request for $50,000 to market the rock in return for fifty per cent of the company — with two caveats. The first was that he wanted a ten per cent royalty until he got his investment back, and the second was that he get a northern fishing trip. They shook hands on the deal.

"When Arlene put on the gneiss necklace she said, 'It's really cool having something around your neck that is so old. It gives you a good perspective on how short a time we're here.' She called it good karma. She's an absolutely beautiful person all the way around."

When it was over, Mark told the host, "This was a dream come true." He phoned his brother from the airport to say he was "walking on clouds." Of course, Mark didn't walk out with a cheque in hand. It remains to be seen how a partnership will actually unfold.

The flight to Edmonton was uneventful, but the plane to Yellowknife "went mechanical." Mark went out for a smoke while the repairs were being made, but upon return found that he'd missed his flight. He spent the night in the Edmonton airport, with his last rock for a pillow. He laughs when he talks about it. That's the thing with Mark. Sober or not, he helps me realize that in many ways, life can be hilarious.

The rock, as he says, gives him a place to stand. That night it also gave him a place to lay his head.

Mark says. "I wasn't looking for the rock in the first place, it found me," and so he believes that its story is bigger than he is.

Since establishing his fledgling company, Mark continues working in the camps to keep his fledgling company afloat. He harvests rock, finds craftspeople to polish, carve and set his unique designs, markets and fills orders through his website.

Bob Dylan wrote, "How does it feel to be without a home? Like a complete unknown, Like a rolling stone." This made me think again about human relationships with rocks. Is it about returning home?

I wondered about people using terms like "being grounded" and intentionally working to feel that way, with yoga, for example. I wondered about all those rock collectors, photographers, preachers, scientists, Elders, ecologists and passionate prospectors who want to tell stories about, and connect with, Earth. Is it all an effort to feel less like a rolling stone?

**Questions for Reflection**

1. What helps you to feel grounded?
2. Try to imagine being Mark Brown as he faced the Dragons. In your life, when have you ever reached for the stars?

In Woodstock, Canadian singer-songwriter Joni Mitchell beautifully called us to our senses:

> We are stardust, billion year old carbon
> We are golden, caught in the devil's bargain
> And we've got to get ourselves back to the garden....

In *A Stone for a Pillow,* Madeleine L'Engle disassembles the word *disaster*:

If we look at the makeup of the word disaster, dis-aster, we see *dis,* which means separation, and *aster* which means star. So dis-aster is separation from the stars. Such separation is disaster indeed. When we are separated from the stars, the sea, each other, we are in danger of being separated from [the Sacred].

We have such difficulty absorbing the magnitude of the vast amount of adaptive information that life employs because our human life span amounts to a tiny fraction of cosmic time, approximately a millionth of one percent. Our great challenge then, in comprehending the universe is to overcome our natural bias that the world has always been the way it appears to us. And that it will always be more or less the same as it is now.

We can begin to appreciate something of the changing nature of the universe when we realize that even our means for sensing the processes of the universe are part of these processes as well. The way we see, the way we hear, the way we feel — each of these senses has been drawn forth and deepened for hundreds of millions of years. We see only because the Earth has long been inventing the sense of sight. And this process is not yet done.

Brian Swimme & Mary Evelyn Tucker in
*The Journey of the Universe*

Chapter 8

# Stone Soup and Other Dreams

**Ask the animals, and they shall teach you;**
**the birds of the air, and they shall instruct you.**
**Speak to the Earth and it shall teach you.**

Job 12:78 *The Bible*

In her book, *Wild Stone Heart,* Saskatchewan writer Sharon Butala writes about a field on her ranch. Unsuitable for cultivation or grazing, the field was left alone. It is in this wild state that it reveals its stories to the patient and curious author. Butala traces her relationship with this field in a personal and provocative story that transforms her over a period of twenty years.

Like Thomas Berry, David Suzuki, Anna Marie Sewell, Brian Swimme, Audrey Whitson, Carolyn McDade, David Suzuki, Black Elk, Schim Shimmlar, Butala gives voice to the land. Frank Powderface in *Stoney History Notes*, invites readers to "listen to the stories hidden in this landscape."

We need to learn to listen. We know that all the trinkets and toys of modern life, whether "smart bombs," or the latest five hundred dollar purse are an effort not to listen. We are worse than those who block their ears and sing *lalalala I can't hear you.* We know we need to stop rolling, start listening and find our way home. Ancient stories might hold clues to help us.

### The Lost Child

Once upon a time there was a man who had two sons. The younger one said, "Father, give me my share of the estate. I don't want to hang around here on the farm any more. I've got plans. Dreams. And so, the father handed over his inheritance and the son went on his way into the great wide world.

And the world was full of music, wild women, fine wines and exotic food. He found a multitude of friends to help him enjoy his wealth and they partied night and day.

When the money ran out, his friends did, too. When he sobered up, he realized that the country was in crisis, suffering a drought and famine. Unless you were rich, times were tough. Eventually, he went to work as a swineherd. Out of sight of anyone but the pigs, he knelt and ate with them.

The story says that eventually "he came to himself" and said, "My father looks after his servants. I'll ask his forgiveness, and beg to become his servant."

When he was still a long way off, his father saw the thin, ragged, dirty man walking up the lane. He recognized him, and ran to meet him, tears of joy streaming down his face. "Welcome home, beloved child!" he cried. He called his servants to prepare a feast, placed a ring on his finger, and dressed him in a fine robe. "For my child was dead and is alive again; he was lost and now is found," he declared.

Like prodigal children, many of us are waking up to realize that we are making a "stinking pig pen" to live in — polluted oceans, lakes, rivers, land, air. Although some with money can hide in gated communities and drink bottled water, there is discomfort and fear in that. The cost of this lifestyle is more than mere dollars; it can cost the soul. I believe that most of us, at some level, are wondering how to get home. Back where the spirit feels good and our home is safe from destruction.

The Lost or Prodigal Son story was told by Jesus of Nazareth two thousand years ago. I think it has wisdom for today. Even the idea of homecoming is a touch-stone. Who doesn't want to return home after a hard day's work, a busy life, a great adventure, a frightening decision, a stay in hospital?

Some wonder if the Acasta gneiss could be our touchstone. Can it give us another perspective? Can this rock help us wake up and return home? I've always loved a touchstone in another ancient tale.

**Stone Soup**

When a traveller entered the gates of a village one day, the villagers ran away and locked their doors. They were suspicious of strangers.

He said, "Hello, people! Don't be afraid. I'm just a simple traveler, looking for a simple meal."

"Keep moving or I'll sic the dog on you! There's not a bite to eat in the whole village," shouted a man. "We would share if we had enough, but we don't," shouted a woman. "Go away!"

"Oh, I have everything I need," the stranger said. "In fact, I was thinking of making some stone soup to share with all of *you*." And he proceeded to fill his pot with river water and build a fire where all could see. With great ceremony, he then took a stone from his pocket and dropped it into the steaming water.

Hearing the rumor of food, most villagers had cracked open their windows to watch. As the stranger sniffed the steam and licked his lips in anticipation, hunger and curiosity overcame their fear.

"Ahh," the stranger said loudly, "I do like a tasty stone soup. Of course, stone soup with a carrot, now that is hard to beat."

It so happened that a girl named Kate had planted carrots in her very own garden that year. She pulled up three and went to offer them to the stranger.

After watching Kate share her carrots, a man said, "I have one potato. Would a potato be good in stone soup?"

"Oh yes!" said the stranger. A woman said, "I could add a bit of beef." The soup simmered. The stranger stirred. The village people licked their lips.

Soon all the village people wanted to share what little they had; beans and bok choy, lentils and cabbage. The soup simmered. The stranger stirred. The village people licked their lips.

Finally, the stone soup was ready. The village people ran for bowls and spoons and Kate, the girl who had started all that sharing, ladled the soup to each person. It was a grand feast! Afterward, the village people brought out drums and lutes, banjos and harmonicas and made music. The children joined hands and danced around the fire until the first stars appeared.

The next morning, the mayor offered the stranger money for his magic stone, but the traveller said it was not for sale. The children followed the stranger to the edge of the village. "Goodbye!" they called. "Come again!" The stranger gave his stone to Kate. "This stone is only magic," he whispered, "because it brings out the magic in people." As children are clever beings, Kate smiled and said, "I figured."

Stone Soup makes me laugh when I consider the men who hold claims on the Acasta gneiss. On one hand they say that they'd like to make money from it. On the other hand, they have given it away freely.

When he held a claim, Walt Humphries gave it to museums and universities; Jack Walker has given tons of it to the Smithsonian Institute, and when I interviewed him he gave me a beautiful polished piece. Mark admits that he's given away more than he's sold. In fact, when he auditioned to appear on the Dragon's Den, he gave a piece to the producer. "You shouldn't be giving it away!" she admonished, putting it in her purse. Mark just laughed.

I enjoy the wonderful irony in this. Working with this rock has allowed me to focus on many layers of the Acasta gneiss story. I realized early in my research that I needed to know more than geology to write about that rock. And so I had taken a piece to a wise woman.

**Cree Elder Doreen Spence**

Doreen's home in downtown Calgary is filled with beautiful mementoes of her travels around the world, and many beautiful and exotic rocks. Sacred rocks are kept in pretty beaded, handworked leather or fabric bags. One rock has been used for years in healing ceremonies, another is one she uses for meditation.

Her work as a healer and teacher takes her frequently to Europe, Mexico, Central America, Australia, South Africa and throughout North America; some of her rocks are from these places.

Doreen is the recipient of numerous humanitarian and peace awards and in 1999 she was named one of the Thousand Women for Peace nominated for the Nobel Peace Prize. At seventy-five, Doreen is serene yet compelling, a gifted story teller and generous in her teaching. She continues to lead vision quests, conduct sweat lodges and fasts. Doreen worked for decades as a nurse; healing has been her life's calling. She herself is full of vibrant health. Her long black braid does not betray her age, nor does her smooth skin. She seems both youthful, and wise.

When I gave her the rock, Doreen said that she felt a band around her forehead. She felt the rock exuded a warm, strong energy. A feminine energy.

"The rock people," she began, "are the wisdom holders, they are the librarians of the universe, the skeletal system for Mother Earth. We get a lot of our messages, our insights and wisdom from the rocks because they have such an energy and knowledge.

"I've had profound experiences with rocks and at first I never shared this with people in the western world. I wouldn't have said, 'That grandfather rock spoke to me or asked me to do something'. It seems more open now; people are more receptive to this way of thinking."

Doreen explained that she was raised in the traditional way by her grandparents. Her grandmother was a medicine woman and her grandfather had the ability to communicate with animals. They left the Goodfish Lake Reserve in northern Alberta to move into the bush in order to save Doreen from attending an Indian Residential School. She was six years old when they spirited her away in the middle of the night.

As a little girl, she says she learned to "listen to the rocks and trees and to learn the different energies. They have their own energy centres within them. They are like us in this way."

Doreen learned to listen to the natural world as a child and her healing work continues this deep listening. "Once I was in Mexico, and alone in the jungle. I knew there was a jaguar nearby; I could smell and feel him. But I didn't feel afraid. I walked along a path and got to a place where there were huge boulders, nine feet tall, about five feet thick and three feet wide. As I came upon them I could feel my third eye buzzing. I kept walking. I can't explain this, perhaps it was an out-of-body experience, but I actually walked through that rock. I travelled through time and space and I found myself with [different] entities. I was told later that they are called *Palaedians*. They welcomed me and were surprised that I was not afraid.

"They didn't relate to me why I was there, but said that they would gift me with a rock. It was hot like fire, it had such energy. This Acasta rock reminds me of that one.

"I thanked them and then went whirling until I was sitting on the other side of the boulder from where I'd started. I had that rock in my hand and knew that it was sacred. I put it in my sling pack. I was having a hard time getting back into my body. I felt like I was still spinning.

"To settle down, I tried the pipe, but nothing I tried helped me focus. I didn't feel part of the planet, but like a little star floating around up there. And I lost the little sacred rock.

"Four days later I was with my friends. I reached into my purse and found my rock again. When I touched it, it felt *hot.* I showed it to my friend and said, 'You read rocks. You do psychometry. Tell me about this rock.' When she picked it up she said it burned her hand. She returned it to me and said, "It's not from this planet."

"Eventually the rock said to me that we are in a time of major changes on this planet. We humans have been experiencing masculine energies — wars — things out of control. But in 2012 we'll see a lot of disasters that will cleanse and purify. We will begin to return to the feminine energy of unconditional love. This [upheaval] is what it will take for us humans to heal one another and Mother Earth.

"I had seen in a vision that Mother Earth would purify herself and new energies would evolve. I've been experiencing these newfound, gentle energies.

"Those of us who are doing the work of unconditional love, are experiencing far deeper, more profound feelings of compassion for one another. It's like osmosis. Because Mother Earth is experiencing it, we are also. That's what I get from this rock, the Acasta Rock. I haven't spoken to it, but I get the same feeling from it, I get warm, hot energy coming from it. If you can feel energy, that's what you'll feel from it."

I told Doreen that when I first began writing this book I had a sudden desire to weep because I felt such overwhelming compassion for the planet. That was when I knew I wanted to bring the rock to her and to sit with her. She said, "Home. It's like a shawl when you wrap it around you; it's a feeling of home. With a rock, when you find them or they find you, [you may weep] because you are reconnecting with a long lost relative."

Doreen reminded me that rocks have their own healing energies and vibrations. "That's why we use the rock people when we go into the sweat lodge. We use the rocks because they know how to take away the pain, anger, the emotional stuff that a person carries that doesn't serve their highest good. Nothing is stronger than a rock for this high purpose. They are the ultimate grandmothers and grandfathers."

Doreen ended our session with another story, this one about a child. Her granddaughter was only three when her parents separated.

Doreen said she used the rocks to teach. "We'd sit on my bed with the rocks. She chose one and I made a little pouch so she could keep it with her. It was comforting and she kept it for a long time.

"But one day, sobbing her heart out, she told me that someone had stolen her rock. I said, 'whoever took that rock person must have really needed it. Maybe their mother was sick or something bad has happened in that person's family. Something caused that child to take your rock. Obviously the rock person needed to go with that person, otherwise it would never have left you. So, now you have to pick another rock and this one will help you with the next stage. You have dealt with certain things with that rock, now it's time to move to the next level.

"I put all the rocks on the bed and told her to take her time and choose another. I left her alone. She sat with them for a very long time. Eventually she came bounding out of the bedroom and said, "Grandma! I found it! It feels good!" She'd chosen a purple amethyst. An amethyst is for cleansing and purification. You see, children know."

As a child, Mark had always felt a part of the natural world, as I had. The farm he lived on near Orangeville, Ontario is in the rolling Caledon Hills. This is maple syrup, corn and dairy country. It's forest and pasture, small towns and small farms. His bedroom was always filled with bones, porcupine quills, rocks, fossils, snake skins, bits of wood, abandoned hornet's nests. Floor to ceiling, covering book shelves, peeking out of boxes — it was easy to find some part of the countryside moved into his room for closer inspection or for the simple tactile pleasure that a piece of paper-thin birch bark or sun-bleached fox skull gives.

Mark said that he didn't go out searching for this rock, but that the rock seems to have found him. Will his discovery of it chart a future path for him?

"I can go out on some weird trips," he told me, "but when I put my hands in my pockets and touch the gneiss, I can find my way home.

"People say diamonds are forever. They're not. The gneiss is forever. When the royal couple came here, the government gave them diamonds as a token of thanks. They are valuable and all, but

I've worked mining exploration; there are diamonds *everywhere*. That *forever* business is just de Boers marketing."

And, as Jack Walker has said, this rock is not about mining or gems. It's something else. As I speak with people about this rock, I wonder. Is this our global Stone of Destiny? Is it a symbol to help us turn away from the environmentally destructive path we are on? Has this grandfather rock kept his story until this moment in time for a reason?

**Jacob's Dream**

Another ancient story, told by Jews, Christians and Muslims, is about a man who wrestled with the Holy.

Jacob was a tortured soul because he had deceived his ailing father and cruelly cheated his older brother of his birthright. Because of this, he went on the lam. Crossing the wilderness alone, he camped for the night. It must have been scary out there; it often is scary in those wild, in-between places we find ourselves in. At sun set, he lay his head on a "pillow of stone" and had a dream that people still talk about 3,000 years later.

Jacob dreamed of a ladder rising up to the very heavens. It was brilliant, this ladder, and on it were angels, coming and going, an infinite stream. In the morning, Jacob set up his dreaming rock and named it *Bethel* which means, God is in this place. The story seems to let us know that the divine is not up in the sky, but here, on Earth, that we're connected. As I understand it, it gives assurance that Heaven and Earth are one.

That's not the end of Jacob's story or of his connection with rocks. Jacob's story is filled with trickery, intrigue, danger and some humour from beginning to end. He worked for his Uncle Laban, married Rachel and Leah, had many children and became a successful shepherd and rancher. After twenty years, though, he felt compelled to return home — and to confront his brother's wrath. When he set out, he incurred his Uncle Laban's wrath, too. He was good at making people angry, it seems.

When Laban caught up with him, they didn't fight after all. In gratitude, Jacob set up a stone pillar to mark their peace agreement. His family piled stones beside it. Laban said, "These stones bear witness to our promise." I very much like that idea.

*En route* to his brother, Jacob leaves his family and servants and enters into a dreamtime wrestling match with a stranger. Perhaps it is an angel, maybe it's God, or his own conscience; that detail remains shadowy. The battle lasted all night. Although his hip was badly injured, Jacob did not give up the fight until he had wrestled a blessing from his adversary. That's what he wanted more than anything in the world, a blessing and peace of mind.

Over dinner one night I asked Bill about Jacob's wrestling match. Bill served as the Moderator (spiritual leader) of the United Church of Canada, is a retired minister and former lawyer. He was thoughtful. "One thing I take from the story is that we aren't wrestling with spiritual questions much these days," he said. Bill believes that the big questions *are* spiritual: how do we treat the Earth and her creatures? How do we manage our economy? How do we work together for the common good of all?

If we were to lay our heads or simply rest our minds on the Acasta gneiss, would it give us courage enough to wrestle with these spiritual questions?

Humility is a word related to humus, soil. Bill describes a lesson in humility in *Cause for Hope: Humanity at the Crossroads.* "As I gazed over the flat, endless tundra of the North, all I saw was a brownish grey wasteland. Wilderness. Nothing there. It was only when I dropped to my knees and looked closely at the ground that I saw the delicately coloured flowers, the multi-shaded mosses, the crawling insects, a bird's nest. The tundra was literally teeming with life. But it was only when I knelt in an attitude of prayer that I realized where I was.

"The whole experience humbled me, challenged my assumptions, put me in a new place. Humility is the necessary centre from which we can dare to create a New Story [of care for Earth]. Humility means recognizing our small yet unique place in all creation. Humility means understanding how limited our knowledge and experience really are."

What if, a friend recently mused, we were to stand on that rock — a world summit of sorts — and discuss our common future? It could not happen that way, of course, but considering that there are pieces of the gneiss in museums, universities and collections around the world, it is possible to imagine the rock drawing us into a sacred Circle of healing for Earth.

Part of the Acasta gneiss outcrop is embedded deep in the Earth, part of it stands bold against the wild sky and part of it swoops down into the Acasta River. It connects, therefore, Earth, air, water. It remains to be seen what we will make of it.

**Questions for Reflection**

1. The author believes that ancient stories, such as The Prodigal Child, Stone Soup and Jacob's Dream hold wisdom for today. Do you agree? Why?
2. Bill Phipps wrote, "Humility is the necessary centre from which we can dare to create a New Story." What leaders today do you think have both humility and a vision for the future?

> Our society is one where it's a given that we're greedy and destructive. The great work of our time is building the new and vibrant Earth Community. We'll find our way forward by feeling our way, and then thinking about the feeling.
>
> Brian Swimme in
> a lecture in Calgary, Alberta June 2012

# Afterword

This rock, as I said in the Introduction, has led me on a wild and wonderful exploration of planet Home. For me, this humble rock sparkles not with riches or gemstones, but with the very story of its life. It drew me into its own magical circle, simply by existing.

The Earth Charter, which began as an initiative of the United Nations, offers us a "vision of hope and a call to action" and urges us to think holistically, globally and with compassion for each other and for our home.

The Charter was completed by a global network of citizens concerned about Earth sustainability. It was launched as "The Peoples' Charter" in the year 2000. It begins, "We stand at a critical moment in Earth's history, a time when humanity must choose its future."

For me, the Acasta gneiss offers a vision of hope because it represents the foundation of everything. It reminds me that we are but a small part of the whole Earth, indeed, the whole universe. At the same time, we are a critical part of the whole because we are capable of such thoughtlessness, violence and destruction.

Of course, we are also capable of thoughtfulness, great brilliance and healing ways. We must choose. If our decisions lead to life, then generations will live to marvel at the great age of the Acasta gneiss. Otherwise, the ancient rock will remain, but only the wind will tell its story.

There are millions of reasons for optimism. Citizens of all ages are rising up to lead our leaders, and becoming ever more creative in forming new networks and vibrant organizations at both the local and international levels.

The Earth Charter concludes, "Let ours be a time for the awakening of a new reverence for life, the firm resolve to achieve sustainability, the quickening of the struggle for justice and peace, and the joyful celebration of life."

May it be so.

Acasta River, NWT

**Questions for Reflection**

1. Are you familiar with the Earth Charter?
2. How might you help to make the Earth Charter principles more a part of social discourse and action?
3. In what way has the story of the Acasta River gneiss touched you?

## if you must touch river rock

speckle-backed salmon
fists of ochre
iron footprints
all laid bare by your caress

and here was *Iktomi,* the spider
when this world ran hot and new
and could be woven
she laid lines into this stone then
lines which you uncover
as you smooth back time and wash layers away
until the old ones sing
that young world song

put it in your bag for remembrance

and if making a medicine bag
retrieve as well a blood red stone
for courage
however worn and tumbled
it is red to the core

receive a yellow one
for the sun
a green one for the light that ties
one heart to another, binding wounds
blue for the moon, black one for midnight
a pink one to speak of forgiving
and a piebald one as evidence
that all of us are two forces meeting
which give rise to the child we are

let some stones be porous
some rough, some smooth
so your fingers are soothed
and learn to touch like water
and if one is grey

call it writer's stone
all colours or none
heavy or light
the fog that conceals
or clouds the birthplace of rain
the mind of the poet, story of it hidden
let this one remind you
the beating of a heart
much older than our own
enfolds us our blood
part of that dancing
over stones
that give the river its song.

Anna Marie Sewell

Snowy Owl

## A Place to Stand: a Workshop

Approximate time: 2.5 hours

**Purpose:**

1. to invite participants to contemplate their personal connection with Earth
2. to invite participants to consider their place in the ongoing story of Earth

### Ahead of Time

1. Invite participants to bring a rock, gem or stone from home to the workshop.

2. Invite participants to read The Covenant and Principles for Childhonouring.
   http://www.childhonouring.org/covenantprinciples.html

3. Invite participants to read The Earth Charter
   http://www.earthcharterinaction.org/content/pages/Read-the-Charter.html

4. Invite participants to view The Acasta River Gneiss YouTube video:
   www.youtube.com/watch?v=pJTBCeVGPIQ

5. Print the closing words on a large sheet of newsprint to be read in unison (or prepare to project the words.)

**You will Need:**

1. a printed or projected copy of The Earth Charter

2. a printed or projected copy of A Covenant for Honouring Children

3. a reading from a book such as *Earth Prayers* by E. Roberts and E. Amidon (or use the one provided)

4. refreshments.

5. *Option*: Set up a book and video table with titles gleaned from the Bibliography of this book, your own collection, or ones from the public library.

6. *Option:* Find out if a piece of the Acasta River gneiss is on display at a museum, university or public collection in your community. Suggest a visit.

7. *Option:* Sing or listen to Pat Mayberry's song, *Called by Earth and Sky.*

## Room set up:

Arrange chairs in a circle. Provide a low table in the centre. Place a candle and matches, a small bowl of water and a feather on the centre table. When participants arrive, invite them to place their stone on the table, too. The elements, Earth, air, water, fire are thus represented as a focus.

## The Workshop:

Welcome participants. To begin, invite a volunteer to light the candle and to read aloud a short reading you have selected or read from Job 12:78:

"Ask the animals, and they shall teach you; the birds of the air, and they shall instruct you. Speak to the Earth and it shall teach you."

*Invite* everyone to state their name and tell of their happiest memory of being in Nature when they were a child. Ask people to try to include a thought about how Earth "taught" them something.

*Ask,* What ecological problem, large or small, are you concerned about? *Option:* If you see that there are groupings of similar problems, you may wish to invite participants into small groups to share ideas and/or discuss solutions. Participants may report back or not; use your descretion.

*Invite* participants to introduce the rock, stone or gem they have brought: Where is it from? What kind is it? How long have they had it? Why is it significant to them?

*Summarize* in your own words what the Earth Charter says and say how you personally feel about the Charter.

Passing a copy around the Circle, *invite* participants to read one paragraph in turn from The Covenant for Honouring Children.

*Ask*, What stands out for you in the Covenant and the Charter?

*Invite reflection* about how the Covenant and the Charter intersect and compliment one another.

*Invite* participants to brainstorm: Who are the leaders in the community or in the world who seem to have the values espoused in the Charter and the Covenant? After naming them, suggest writing to them to encourage their vital work.

*Ask:* How do the rocks, stones and gems we cherish give us a place to stand as we face the ecological emergencies of our time?

## Closing

*Ask,* As a group, is there further action for Earth that we might take? If so, plan the next step.

*Invite* closing reflections from participants.

Read in unison the final words from The Earth Charter:

**Let ours be a time remembered for the awakening of a new reverence for life, the firm resolve to achieve sustainability, the quickening of the struggle for justice and peace, and the joyful celebration of life.**

# Thank you for the Map and Photographs

Thank you to Brock University for the use of the map of Northern Canada. Northern Canada Outline Map. Brock University Map Library. Controlled Access:
www.brocku.ca/maplibrary/maps/outline (Accessed May 13, 2013)

Thank you to Stefan of Whitehorse for use of the golfing photographs. The Acasta caribou antler photograph and back cover photo of Mark harvesting rock are by Jeremy Emerson.

Grateful thanks to the Creative Commons photographers for interior photos. Originally, all these beautiful images were in colour. Jeremy Emerson who kindly supplied the photo of the caribou antlers on the shore of the Acasta River on page 6, is not a Creative Commons photographer.

| | |
|---|---|
| Caribou | Dave Menke |
| Arctic Hare | U.S. Fish & Wildlife |
| Raven | Doug Brown |
| Arctic tern eggs | Mike Beauregard |
| Mountain Avens | Anne Burgess |
| Owl | Floyd Davidson |
| Arctic Wolf | xploitme |

For information about photographs from the Creative Commons, or to see these wonderful pictures in colour, please visit: http://creativecommons.org/licenses/by-sa/3.0/

## Bibliography

Barnes-Svarney, Patricia & Svarney, Thomas E. ***The Handy Geology Answer Book*** Visible Ink Detroit 2004

Bastedo, Jamie ***Shield Country: LIfe and Times of the Oldest PIece of the Planet***

1994 Komatik Press, Arctic Institute of North America, University of Calgary Republished by Red Deer Press 2002

Bjornerud, Marcia ***Reading the Rocks: The Autobiography of the Earth***

Westview Press 2005, Perseus, Cambridge MA

Blondin, George ***When the World was New*** Outcrop Publishing Yellowknife 1990

Brukoff, Barry and Neruda, Pablo ***Machu Picchu*** Editorial Sudamericana 2000

Cadulto, Michael J. & Bruchac, Joseph ***Keepers of the Earth*** Fifth House Publishers Saskatoon 1991

Dillard, Annie ***Teaching a Stone to Talk: Expeditions and Encounters*** HarperPerennial 1982

Dowd, Michael ***Earthspirit: A Handbook for Nurturing an Ecological Christianity*** Twenty-third Publications Mystic, Conn. 1991

Eyles, Nick ***Canadian Shield: The Rocks that Made Canada***2011Fitzhenry and Whiteside, Markham, Ontario

Findley, Timothy ***From Stone Orchard: A Collection of Memories*** Perennial Canada 1998

Flannery, Tim ***Here on Earth: A Natural History of the Planet*** by HarperCollins Publishers 2010, Toronto, Ontario

Fumoleau, Rene et al ***Denendeh: A Dene Celebration*** The Dene Nation 1984

Gray, Elizabeth Dodson ***Green Paradise Lost: Remything Genesis*** 1981

Hitchon, Brian, ed ***Alberta Beneath our Feet: The Story of Our Rocks and Fossils*** GeoScience Publishing 2006

Jones, Adrian P. ***Rocks and Minerals*** Harper Collins 2000

Kostash, Myrna & Burton, Duane ***Reading the River: A Traveller's Companion to the North Saskatchewan River*** Coteau Books 2005

Laird, Ross A. ***A Stone's Throw: The Enduring Nature of Myth*** page 44-45 McClelland and Stewart 2003

Lamb, Simon & Sington, David ***Earth Story: The Shaping of our World*** 1998 BBC Books, London, England (to accompany BBC tv series)

Little, James, editor ***Way Out There: The Best of Explore*** GreyStone Book Vancouver 2006

Littleton, C. Scott, ed. ***Mythology: The Illustrated Anthology of World Myth and Storytelling*** Duncan Baird Publishers, London 2002

Lovelock, James ***The Ages of Gaia: A Biography of our Living Earth*** Norton Publishing 1988

Mehl-Madrona, Lewis ***Coyote Wisdom: The Power of Story in Healing*** Bear and Company Rochester 2005

Meuninck, Jim ***Medicinal Plants of North America*** FalconGuides, Globe Pequot Press Guilford, CN 2008

Murray, Joan ***Rocks: Franklin Carmichael, Arthur Lismer and the Group of Seven*** McArthur & Co. Toronto 2006

Peilou, E.C. ***A Naturalist's Guide to the Arctic*** University of Chicago 1994

Pfister, Marcus ***Milo and the Magical Stones*** North-South Books, New York 1997

Phipps, Bill ***Cause for Hope: Humanity at the Crossroads*** CopperHouse Press Kelowna 2007

Pogue, Carolyn Czarnecki ***Yellowknife*** Outcrop Publishing, Yellowknife, 1981 and 1984

Richter, Joanne and Coutts, Ian ***Canada's Rocks and Minerals*** (YA) Scholastic Toronto 2007

Schaef, Anne Wilson ***Native Wisdom for White Minds: Daily Reflections Inspired by the Native Peoples of the World*** One World Book/Ballantyne 1995

Sutcliffe, Antony ***On the Track of Ice Age Mammals*** Harvard University Press Cambridge Mass 1985

Swan, James A. ***Sacred Places: How the Living Earth Seeks our Friendship*** Bear and Company New Mexico 1990

Swimme, Brian and Mary Evelyn Tucker ***Journey of the Universe*** Yale University Press 2011

Taylor, Jowi ***Six String Nation*** Douglas & McIntyre Vancouver 2009

Turk, Jon ***The Raven's Gift: A Scientist, a Shaman, and Their Remarkable Journey Through the Siberian Wilderness*** St. Martin's Griffin, New York 2009

van Herk, A. ***The Tent Peg*** McClellend and Stewart 1983

Whitson, Audrey J. ***Teaching Places*** Wilfred Laurier University Press 2003

Wilson, John ***Dancing Elephants and Floating Continents: The Story of Canada Beneath our Feet*** Key Porter 2003

Wilson, John & Clowes, Ron ***Ghost Mountains and Vanished Oceans: North America from Birth to Middle Age*** Key Porter Toronto 2009

***The Holy Bible*** The Books of Genesis, Job and Luke

**Videography/Websites**

***Rock of Ages, NWT*** www.rockofagesnwt.com
***The Acasta River Gneiss***
www.youtube.com/watch?v=pJTBCeVGPIQ
*****The Awakening Universe: A Liberating New Cosmology for our Time*** a film by Neal Rogin/ Brian Swimme interview
***Charles Darwin and the Tree of Life*** BBC Earth 2009 hosted by David Attenborough
***Planet Earth: As you've never seen it before*** 2006 BBC Discovery Channel and CBC DVD
***Yellowstone: A Battle for Life*** BBC America DVD 2009
***How Earth Was Made*** *Pioneer TV, History Television 2009*
**Creation Hymn**:
http://www.mountainman.com.au/news97_8.html
**Sleeping Crow** - Are rocks alive?
http://www.scribd.com/doc/57336927/Red-Tao

**Interviews:**

Mark Brown 01/12, 06/13, 06/12 & 05/13
Walter Humphries Yellowknife, May, 2012
Jack Walker Calgary April, 2012
Jeremy Emerson Yellowknife May 2012
Wouter Bleeker phone March, 2012
Fred Sangris (with Aggie Brockman) Yellowknife 06/2012
Janet King phone June 21, 2012
Gina Marie Ceylan phone June 27, 2012

## Index